# The Evolution of Innovation Networks

Tobias Buchmann

# The Evolution
# of Innovation Networks

## An Automotive Case Study

With a foreword by Prof. Dr. Andreas Pyka

Springer Gabler

Tobias Buchmann
Lehrstuhl für Innovationsökonomik
Universität Hohenheim
Stuttgart, Germany

Dissertation, University of Hohenheim, 2014

D 100

ISBN 978-3-658-10382-8      ISBN 978-3-658-10383-5 (eBook)
DOI 10.1007/978-3-658-10383-5

Library of Congress Control Number: 2015942135

Springer Gabler
© Springer Fachmedien Wiesbaden 2015

Printed on acid-free paper

Springer Gabler is a brand of Springer Fachmedien Wiesbaden
Springer Fachmedien Wiesbaden is part of Springer Science+Business Media
(www.springer.com)

## Foreword

Innovation networks are a complex organisation form of industrial R&D which plays a prominent role in the generation and diffusion of new knowledge. Economists widely ignored innovation networks and claimed that the phenomenon must be transitory only, a consequence of disruptive technical change that will disappear. Instead of trying to understand the dynamics of knowledge development, the whole phenomenon was reduced to the discussion of spillover effects which are likely to distort the incentives of firm actors to invest costly resources in research and development activities.

Because of the sheer existence of this cooperative form of industrial research, innovation networks are considered an expression of exceptional circumstances. Innovation networks might allow established firms to get access to relevant but distant knowledge introduced by innovative start-ups. After the creation of own competences in the respective fields, the innovation networks disappear and with them the small start-up companies. If this is not the case and the established companies are not able to integrate the new competences, the start-ups will become the established firms of the future, replacing the old establishment.

Economic development has shown that traditional economists were barking up the wrong tree: In many industries, innovation networks are not a transitory but a permanent phenomenon which connect heterogeneous firms in their attempts to improve the knowledge base. It is the merit of Tobias Buchmann that he addresses this important topic not only from a new theoretical perspective embedded in Neo-Schumpeterian economics but also complementing his theoretical reasoning with an important and so far rare empirical study.

In his thesis, he has outlined a conceptual framework for capturing network evolution patterns of interfirm innovation networks and analysed the dynamic evolution of an R&D network in the German automotive industry. In particular, he tested a number of hypotheses with respect to the drivers of evolutionary change processes of a network that are based on subsidised R&D projects in a recent period encompassing ten years.

For this purpose, he employed a stochastic actor-based model in order to estimate the impact of network change drivers. In his analysis, which can be characterized as a pilot study in understanding network dynamics he is able to derive interesting results. For example, he showed that structural positions of firms as well as actor covariates and dyadic covariates are determinants of the evolution process.

Tobias Buchmann's result are likely to meet a large interest in the field of modern innovation research. The results are relevant to support the strategic considerations of firms involved in networks as well as of policy makers in the field of innovation policies.

Prof. Dr. Andreas Pyka

## Acknowledgements

I would like to thank my first supervisor Professor Pyka for his support, advice, useful suggestions and for the freedom he granted me to work on the dissertation and my research projects beyond. I thank my colleagues Benjamin Schön, Matthias Müller, Dominik Hartmann, Malcolm Yadack, Ben Vermeulen, Sophie Urmetzer, Bianca Janic and my co-authors Muhamed Kudic and Daniel Hain with whom I had inspiring discussions and a good time. Furthermore, I am very grateful to the Friedrich Ebert Stiftung (FES) which granted me a PhD scholarship. The fruitful and inspiring discussions with other scholarship students during seminars as well as the financial resources had a significant impact on the successful completion of my dissertation project. In this regard, I also thank Professor Hagemann for his support during the scholarship application procedure and for his willingness to co-supervise my dissertation. Special thanks go to my parents whose encouragement and support gave me the chance to attend university studies. Last but not least, I thank my wife Judith and my daughter Milla for their patience and support.

Tobias Buchmann

# Table of Contents

# List of Figures

# List of Tables

# List of Abbreviations

| | |
|---|---|
| ABM | Agent-based modeling |
| BMBF | Bundesministerium für Bildung und Forschung |
| CAN | Controller area network |
| CAS | Complex adaptive systems |
| EPO | European Patent Office |
| ERGM | Exponential random graph model |
| GDP | Gross domestic product |
| GNSS | Global navigation satellite system |
| ICT | Information and communication technology |
| IPC | International patent classification |
| IPR | Intellectual property right |
| JPO | Japan Patent Office |
| KBR | Knowledge-based resources |
| MCMC | Markov Chain Monte Carlo |
| MQB | Modularer Querbaukasten |
| NACE | Nomenclature statistique des activités économiques dans la Communauté Européenne |
| NCRA-RJV | National Cooperative Research Act-Research Joint Venture |
| OECD | Organization for economic cooperation and development |
| OEM | Original equipment manufacturer |
| OLED | Organic light-emitting diode |
| OLS | Ordinary least square |
| PATSTAT | EPO Worldwide Patent Statistical Database |
| PBR | Property-based resources |
| PCT | Patent cooperation treaty |
| PMF | Probability mass function |
| QAP | Quadratic Assignment Procedure |
| R&D | Research and development |
| SAOM | Stochastic actor-oriented model |
| SME | Small and medium enterprises |
| SNA | Social network analysis |
| TFP | Total factor productivity |
| TV | Television |
| US | United States |
| USPTO | United States Patent and Trademark Office |
| WIPO | World Intellectual Property Organization |

*"Model building
is the art of selecting those aspects of a process
that are relevant to the question being asked.
As with any art, this selection is guided by the
taste, elegance, and metaphor;
it is a matter of induction,
rather than deduction. "*
(Holland, 1995, p. 146).

# 1. Introduction

## 1.1 Complexity – Holistic and Reductionist Views in Economics and Cuisine

Intellectual nourishment is a typical and well-accepted diet for doctoral students. However, the growling of the stomach is often an unambiguous signal that food of a more solid variety is called for. Following this call, you may pass by a supermarket on your way home from the office. Myself, always appreciative of *la cucina italiana*, I often decide to grab some spaghetti, a string bag of garlic and onions, a cup of cream and some ham. If you have olive oil and basil at home, all ingredients are kept at hand to prepare a simple but nice spaghetti dish with a ham-cream sauce. As you would expect, this dish is delicious and can be well complemented by a glass of a characterful *Chianti Classico* wine. Now you can simply enjoy the delicious food and wine, or you may be in the mood for scientific inquiry and ask the following question: How do these simple clear-cut ingredients transform into a delicious dish with a complex and multifaceted flavor?

To answer this question, we may start with the analysis of the dish. A cooked dish can be regarded as a typical example of a complex system. Simon (1995, p. 26) considers "a system as complex if it can be analyzed into many components having relatively many relations among them, so that the behavior of each component depends on the behavior of others". To understand the functioning and outcome of an entire system (the dish), it is not sufficient to only analyze the constituent parts (the ingredients, the molecules, etc.) in isolation, i.e. separated from each other. In fact, cooking essentially means that we mix the ingredients according to a (mental) recipe and heat them up. When we mix the garlic, onions, cream and ham, and when we add some olive oil, basil, salt and pepper, then we hope that the resulting sauce will not taste like a bit of ham, a bit of garlic, a bit of basil and so on, but we hope that something new, tasty and maybe surprising emerges from the cooked mix of well specified ingredients. Indeed, the described sauce develops a specific character with a flavor and taste that can hardly be anticipated from tasting the ingredients separately. In other words, from degusting the ingredients separately we can hardly predict the exact taste of the prepared dish. Furthermore, while it is nowadays easy for natural scientists to study the constituent parts of a dish down to its smallest components, it is much more challenging to re-engineer food based on its components and understand the interaction patterns of components. Ahn et al. (2011) give it a try in attempting to understand food preferences as a function of the ingredient mix. They study a large number (56498) of recipes to find general patterns which explain ingredient combinations in food. Based on a network approach which links ingredients if they share at least one flavor compound, they develop a flavor network. Their hypothesis states that our food is more frequently based on ingredient pairs that are strongly linked in the flavor network. The results in fact indicate that the hypothesis holds for North American and Western European cuisines, but that East Asian and Southern European cuisines do not show a preference for recipes whose ingredients share flavor compounds.

Two lessons can be learned from this little excursion into the culinary world: First, a doctoral student can hardly escape from thinking of his dissertation project even when he is cooking. Second, as Aristotle already pointed out: "In the case of all things which have several parts and in which the totality is not, as it were, a mere heap, but the whole is something beside the parts [...]" (Aristotle and Ross, 1953, p. 129). More recently, Chen (2008, p. 91) puts this phenomenon in similar words: "Both atomism and reductionism assume that the whole is the sum of all the parts. When complex interactions exist between different elements, the whole is more than the sum of parts, especially in living organism and social organization."

Complex interaction between the constituent parts plays a crucial role for the emergence of a greater whole. This holds for the ingredients of a dish as well as for individuals or organizations in socio-economic systems. Based on this understanding more specific questions can be asked: In which way are the parts tied together? How do they interact in that they create the observed (or rather tasted) result? Besides food, the lessons learned and the questions asked can be transferred to less sensory-oriented contexts. For example, people acting in a social context are embedded in complex structures and – as analyzed in this dissertation – firms cooperating with the aim to become more innovative are the ingredients of innovation networks or innovation systems. How do these firms interact and what determines their preference for specific cooperation partners? Capturing interaction between heterogeneous agents is a vital step on the path to understanding the emergence and evolution of complex structures.

Networks are evolving and adapting structures steered by the motives, behaviors and objectives of actors (Kash and Rycroft, 2002). Chen (2008, p. 82) argues that "the many-body problem (such as social behavior) is essentially different from the one-body (in a representative agent) and two-body (in bilateral bargaining) problems". The emergence and evolution of social structures resulting from motives of interaction is particularly interesting to study (Arrow, 1994).

For the case of interfirm networks, Powell (1990) suggests in his influential paper with the title "Neither Market nor Hierarchy: Network Forms of Organization" that collaborative ensembles are characterized by lateral or horizontal patterns of exchange, independent flows of resources and reciprocal lines of communication. Moreover, group phenomena are said to influence decisions taken by individual actors, constituting a repercussion effect on micro actors. In networks, actors react and adapt their behavior to the decisions taken by individuals or subgroups of other actors – for instance their neighbors – in the past (social influence). Firms follow strategic objectives aiming at the creation of novelties, the generation of profits, etc. By modeling heterogeneity, social behavior and bounded rationality, individual action turns into complex aggregation processes. Firms are heterogeneous as they have their own specific history, personnel, knowledge-base etc. These firm characteristics (variety) serve as the basis for the development of differentiated industries (Pyka and Fagiolo, 2005). Interfirm networks constitute a specific type of *social network* which can be defined as „a specific set of linkages among a defined set of actors, with the additional property that the characteristics of these linkages as a whole may be used to interpret the social behavior of the actors involved" (Mitchell, 1969, p.2).

Going back to food, it cannot only stop the growling of our stomach and serve as an example for a complex system, but also inspire economists to defining an important concept related to innovation and production processes of firms, namely *technology*. Dosi and Nelson (2010) define technologies by the notion of a recipe which contains the design and the necessary procedures for the creation of a product together with the specification of actions, tools and ingredients (input factors) that have to be combined in a specific manner. However, only parts of a production process can be codified and substantial parts consist of *tacit knowledge*. Tacit are those knowledge elements which are embodied in individuals but differ from individual to individual and cannot be fully codified in words or writing.[1] "Tacitness is a measure of the degree to which we know more than we can tell" (Dosi and Grazzi, 2010, p.176). In addition, tacit knowledge elements can, to a relatively high degree, be shared by people that experience the matter in a similar way (Polanyi, 1967). By conferring the characteristic of tacitness to technological knowledge, it cannot be treated as a pure public good anymore since the replication and development of the necessary capabilities to access and make use of knowledge and technology involve considerable costs and require considerable learning efforts (Dosi and Nelson, 2010). Grant (1996) concludes that explicit knowledge becomes ubiquitous through its communication while the disclosure of tacit knowledge requires its application. Furthermore, relevant knowledge is typically distributed among a greater variety of actors and successful innovation processes are related to timely and locally bound combinations of knowledge. Accordingly, technologies inherently contain social elements linking organizations and their constituent parts. In this perspective, technology has to be analyzed in connection with organizations, social structure and the division of labor. The notion of social technologies is applied to point out that technologies are shaped by norms, beliefs and social practices (Nelson and Sampat, 2001). Therefore, "a technology can be seen as a human designed means for achieving a particular end" (Dosi and Nelson, 2010, p. 55).

To answer the previously raised questions about the interaction of constituent parts of complex systems, plenty of approaches can be considered. The orthodox microfoundational framework, which represents the established "gold standard" in various scientific disciplines, suggests that we need to understand the ever smaller units aggregates are formed of, and that we ought to find uniform laws derived from the micro entities which explain the behavior of the aggregate. A more system-oriented scientist would contrarily assert that it is more promising to understand micro mechanisms of parts and their behavior in an environment which is shaped by an interconnected micro-macro level architecture. The principles of strong reductionism, represented by the former approach, have been prevailing for a long time in many

---

[1] A historic example for the relevance of tacit knowledge for innovation is the attempt of the Prussian king Frederick the Great to copy Watt's atmospheric engine. When the king sent two men to England to spy out Watt's invention, they "made notes and drawings of the engine" (Redlich, 1944, p. 122). Based on the stolen information, in August 1785 the Prussian technicians were able to construct a machine which was employed to drain a copper mine. However, due to technical deficiencies, they were not able to keep the machine in operation for long. Consequently, a second "expedition" was undertaken to gain more information on the functioning of the machine. This time, rather than making more notes and drawings, a special cylinder which was a key element of the machine was brought back to Prussia together with and English artisan (skilled labor), which was illegal at that time. Now the king's engineers were able to build a machine which was more reliable and efficient (Redlich, 1944). Possessing the exact plan of a product or process is not sufficient to create a product or to design a process exactly as it was intended. Instead, parts of the knowledge embodied in products and processes can only be acquired through practical experience (Freeman, 1994).

scientific fields, including economics and management science. However, from understanding the molecules, atoms, protons, electrons, neutrons, quarks etc. of basil, we can hardly guess the taste which emerges when the basil is combined and cooked with the cream and the olive oil. Similarly, we can hardly understand economic aggregates such as innovation networks by analyzing representative firms and research organizations separated from each other. Equilibrium economic models are typically based on the assumption of representative agents. However, the character of the aggregate is often not obvious but emerges from a complex interaction process.[2] Comparable to the example of the ingredient mix which turns into a delicious dish, it is not sufficient to understand the functioning of firms in isolation to explain, for instance, the innovative performance or the emerging structure of an innovation network. Rather, we need tools that enable us to investigate the interaction of heterogeneous firms (Teece and Pisano, 1994) and other relevant actors, because it is the interaction process that shapes the aggregate and in turn, it is the aggregate that influences the behavior of its parts. Consequently, it is indispensable to study interaction in order to understand the functioning of aggregates comprised of ingredients or firms. In particular, "equilibrium assumptions mostly are unwarranted for observations on network processes, and making such assumptions could lead to biased conclusions" (Snijders, 2005, p. 215).

Once we have prepared a dish, i.e. mixed and cooked the ingredients, we cannot restore the garlic, the ham etc. into their original shape. What we have done, that is the cooking, is irreversible. By no means can we extract the ingredients from the dish to have them in their original state. They have been gradually deteriorated and something new has emerged out of them. The Nobel Prize winner Ilya Prigogine called this phenomenon the *irreversibility of time* (Prigogine and Stengers, 1984). It means for a socio-economic system that interaction creates new states and there is no way to reach the exact old state again. Once a (contractual) partnership is formed between two actors, it cannot be dissolved (for a certain period of time). Moreover, the initiated cooperation between two firms changes the network structure but also the actors' characteristics, a process which makes it impossible to reach the previous state again. Economic systems do not show frictionless interaction but are characterized by fluctuations and growth or decline. In this sense, an economic system is more like a

---

[2] According to theories of nonlinear dynamics, every complex system is just an evolving part of an even greater system. Such systems are interleaved with different levels of hierarchy up to the highest level which is the universe as such. "Komplexe Systeme – sowohl chaotische als auch geordnete – sind letzten Endes nicht analysierbar, nicht auf Teile reduzierbar, weil die Teile durch Iteration und Rückkopplung ständig aufeinander zurückwirken" (Briggs and Peat, 1990, p. 221).

biological system than a mechanical system, since economic and biological systems are dissipative in nature and an important characteristic of dissipative systems is irreversibility (Chen, 2008).

## 1.2 Modeling Frameworks

The complexity of systemic structures makes the analysis of evolutionary change often a probabilistic science. Stochastic models are appropriate for this kind of analysis. In this dissertation, the *stochastic actor-based model for network dynamics* (Snijders, 1996; Snijders, 2001; Snijders, 2005) is applied to test hypotheses about the drivers of network evolution of a German automotive innovation network. Studying network evolution refers to understanding the dynamics of the network via some captured mechanisms. In other words, it is about understanding the rules governing the sequence of change through time (Stokman and Doreian, 1997). The applied model cannot definitely prove that a certain driver of network evolution is causal in reality. This can only be shown by collecting and analyzing very detailed observations (e.g. by conducting detailed case studies), which is often impossible for large populations, or by randomized experimental designs. However, agent-based models, such as the applied one, can indeed demonstrate that a certain mechanism is sufficient to explain the observed outcome and to establish plausibility in terms of statistical significance (Holland, 1995).

Mainstream economic theory postulates that aggregate phenomena can be explained by microfoundational theory. But many macro models are based on idealized and rather simplified assumptions with regard to the micro actors, their operational motives and strategies. The usual way to conduct this idealization is by introducing a representative agent (single or categorical agent). An important deficiency of this concept is that individual behavior is scaled up to form the aggregate in a way which assumes homogeneity, homotheticity and identical preferences among all actors (of a category) (Hoover, 2010). Such microfoundational reasoning constitutes an essential building block of neoclassical thinking. Hodgson (1998, p. 169) defines the neoclassical school as follows: "Neoclassical economics [...] may be conveniently defined as an approach which (1) assumes rational, maximizing behavior by agents with given and stable preference functions, (2) focuses on attained, or movements toward, equilibrium states, and (3) excludes chronic information problems (such as uncertainty of the type explored by Frank Knight and John Maynard Keynes)." Such models exclude many aspects which are highly relevant for studying innovation processes as well as innovation network evolution and its drivers.

Approaches for studying microfoundations to understand aggregate outcomes are in general based on the concept of *methodological individualism* as opposed to *methodological collectivism*. The concept of methodological individualism was first delineated in the preface of Menger's (1871) book *Grundsätze der Volkswirtschaftslehre*. The two concepts are described clearly by Samuels (1972, p. 249): "By methodological individualism I mean the view which holds that the meaningful social science knowledge is best or more appropriately derived through the study of individuals; and by methodological collectivism I mean the view which holds that meaningful social science knowledge is best or more appropriately derived through the study of group organizations, forces, processes and/or problems."

Social interaction is in its essential meaning interaction between individuals (Arrow, 1994). To better understand emerging and evolving processes in groups, the strict microfoundational idea needs to be complemented by (i) the study of interaction between actors and (ii) a multi-level perspective considering the influence of aggregates on individuals as a collectivist force. Thereby, the analysis gains a more holistic perspective which helps to better grasp phenomena in social sciences from which we know that individual behavior not only triggers the emergence of group characteristics but the group also influences the individual behavior in a co-evolutionary manner. Clearly, the development of a model always implies that some sort of simplification and thus idealization takes place. However, the developed model must capture essential relationships and mechanisms which are at work in reality and which are relevant for answering the research question at hand. While the neoclassical school stresses generality and precision in conclusions, evolutionary economics, in contrast, emphasizes realism and precision in processes (Van Den Bergh and Gowdy, 2003). By striving for a rather realistic model to study network evolution, I agree with Hoover (2010) that representative agent models do not provide adequate microfoundations for aggregation. Note, that strengthening precision of economic models is rarely a trivial operation. Van Den Bergh and Gowdy (2003) argue that modeling more micro details in parts of a system might go at the expense of losing accuracy in another part of it. Thus, there seems to be a trade-off between realism, precision and generality in economic models (Costanza et al., 1993).

For a long time, there was no real impetus for rocking the foundation of established neoclassical thinking. Despite the weakness in precision and realism, the established theories seemed to explain satisfactorily well important economic phenomena and stylized facts, notably economic growth. Yet, a number of severe problems could not be solved within the established framework. For instance, technological progress as an approved cause for increasing per capita income could not be explained endogenously and not be linked to the variety of innovation strategies of micro actors. Instead, technological progress hit the firms exogenously "like manna from heaven". A contemporary deficiency is related to the global financial and economic crisis which cannot be satisfactorily explained, even though it became a severe threat when the bankruptcy of Lehman Brothers infected the global banking system. Such drastic events in a system are not foreseen in models which oversimplify the microfoundations and neglect deviance from rational behavior. In a different context, collaborative innovation projects have become an effective tool for exchanging knowledge and improving innovative performance of firms in various respects (e.g. time to market, quality, degree of innovativeness), thereby challenging the public good character of knowledge which is a key assumption in many established models. While the collaborative mode of product and process development was barely applied until the early 1980s, the number of collaborative projects sharply increased since then (Hergert and Morris, 1988).

By drawing on the ideas of Schumpeter and on the foundations of the economics of technology and innovation (Nelson and Winter, 1982; Dosi, 1988), Neo-Schumpeterian scholars (e.g. Hanusch and Pyka, 2007a) have attempted to shine more light into the black box of technological progress, evolutionary change and collective innovation. In this school of thought, innovation processes are regarded as the main drivers of economic change and growth. Instead of implementing them as exogenous shocks in models, the Neo-Schumpeterian school explains innovation processes by analyzing their causes, mechanisms and effects. Firms are no longer seen as homogenous entities that maximize their profits by choosing from a set of well-defined alternatives. Rather, firms act within an environment of true uncertainty (Knight, 1921). Moreover, ideas from disciplines outside of economics or management science are welcome in that they provide valuable contributions to the broad repertoire of ideas in Neo-Schumpeterian thinking. In particular, the detailed modeling of micro actors (so called agent-based or actor-based modeling) (Pyka and Fagiolo, 2005) and their interaction patterns requires well defined theories of behavior, connecting economic ideas with psychology and sociology. Not surprisingly, sociological concepts have become a vital cornerstone for the analysis of innovation networks (Burt, 1982; Granovetter, 1985; Coleman, 1988). Smelser and Swedberg (1994) nicely summarize essential differences between so-called mainstream economics (classical and neoclassical economics) and economic sociology: First, the principle of methodological individualism in orthodox microeconomics suggests that actors are not influenced by other actors in a group or society. In contrast, economic sociology explicitly considers actor-interaction relations. Second, economic action is assumed to be virtually always rational in microeconomic theories while rationality is one possible variable value among other forms of action in the latter theory. Third, social structures constrain individual behavior only in economic sociology. For instance, Gulati and Gargiulo (1999) find that sharing a common cooperation partner increases the probability that two firms start to cooperate. In fact, looking beyond the narrow limitations of established economic paradigms of thought is a fruitful way to improve the understanding in many economic fields. The need for scientific openness was already phrased by John Stuart Mill (1865, 2008, p. 62) as: "The only security against [this] narrowness is a liberal mental cultivation, and all it proves is that a person is not likely to be a good political economist who is nothing else."

The development and application of a micro-based network model in this thesis focuses on the study of micro mechanisms of firm interaction. It can be regarded as a possible synthesis of the enduring struggle between individualist and collectivist theories. While in a network perspective it is individual actors making decisions about their collaboration partners (full control about (outgoing) ties), the micro and the aggregate level are not independent from each other but they influence each other, i.e. groups influence individual decisions. A key advantage of a multi-level approach "is that it can incorporate theories that so far have presented opposite, partial and incomplete perspectives on the functioning of macroeconomic systems" (Van Den Bergh and Gowdy, 2003, p. 79). One example is the discussion about selection and social influence. While individual preferences on the one side steer the selection of cooperation partners, it is on the other side the networks which may influence preferences and actor characteristics. The envisaged model should be able to disentangle such effects. Consequently, it becomes increasingly difficult to deliver answers about one-way causalities. We observe instead co-evolutionary processes with complicated interaction patterns. The analysis of firm interaction through the lenses of *Social Network Analysis (SNA)* (e.g. Wasserman and Faust, 1994) enables us to capture actor heterogeneity, e.g. differences in endowments, behavior, learning and absorptive capacities. However, only longitudinal network data can help us to understand the forces and mechanisms of network change over time.

In order to theorize firms, I draw on the resource- and knowledge-based view of the firm (Penrose, 1959; Wernerfelt, 1984; Barney, 1991; Grant, 1996; Pyka and Hanusch, 2006). Knowledge constitutes a key firm resource which is a prerequisite for the development of new products and processes. Early proponents of this concept are Marshall (1920, p. 115) who recognizes knowledge as the decisive factor in production processes ("Knowledge is our most powerful engine of production") and Penrose (1959) from whom we can derive the conceptualization of knowledge as an important resource. A firm's knowledge-base is defined as "the set of information inputs, knowledge and capabilities that inventors draw on when looking for innovative solutions" (Dosi, 1988, p. 1126). Saviotti (2009, p. 27) defines the knowledge-base from a production system perspective as "the collective knowledge that can be used to achieve the firm's productive objectives". In the early 1980s, this approach is taken up by Neo-Schumpeterian scholars (Hodgson, Samuels and Tool, 1994; Dopfer, 2005; Pyka and Fagiolo, 2005; Hanusch and Pyka, 2007a). Here, the role of knowledge for economic development and the success of firms is explicitly recognized and constitutes the cornerstone of economic analysis. Moreover, in the Neo-Schumpeterian perspective networks are seen as a central determinant in the industrial creation of novelty and they are therefore a decisive coordination mechanism. In networks, new technological opportunities are created via technological complementarities and synergies, bringing together a wide spectrum of technological and managerial competencies. Knowledge is no longer considered to be purely a pure public good, but instead as partly local, tacit, firm-specific and complex. These characteristics hamper technological knowledge from being easily exchanged on markets like commodities (Dosi, 1988). Instead, networks serve as an instrument for the exchange and diffusion of knowledge (Valente, 1996; Deroian, 2002).

"The study of networks is part of the general area of science known as complexity theory" (Buchanan, 2003, p. 18). However, Barabási (2005, p. 70) argues that given the current state of network understanding, "network theory is not a proxy for a theory of complexity". While – it is further argued – network theory explains the emergence and evolution of the overall structure representing the "skeleton of a complex system", a more complete understanding requires more knowledge about "the nature of dynamical processes". In the same vein, Newman, Barabási and Watts (2006, p. 7) criticize that "traditional approaches to networks have tended to overlook or oversimplify the relationship between the structural properties of a networked system and its behavior". Learning about such processes implies that we must go beyond the analysis of static structural network characteristics. This means that we need to understand the (generic) micro mechanisms governing the evolutionary change process. With the analysis conducted for this doctoral thesis, I intend to contribute to the understanding of evolutionary network dynamics which can bring us a step closer towards a more complete theory of complexity. Based on a broad set of theories, I derive a number of hypotheses to test the relevance of different mechanisms for innovation network evolution. In particular, I test these hypotheses with empirical data of a publicly subsidized innovation network composed of a sample of German automotive firms.

## 1.3 Research Questions and Outline

A core objective of this dissertation is the development of a model for analyzing the complex evolution of innovation networks and the driving mechanisms underlying network evolution derived from theoretical and empirical findings in economics and related fields. Researchers in economics and management science have devoted considerable effort to investigating the causes, motivations and advantages for the emergence of interfirm networks and strategic alliances (e.g. Powell, Koput and Smith-Doerr, 1996; Ahuja, 2000b; Hagedoorn, 2002). A number of studies focus on motives of cooperative behavior (cf. chapter 4); or they relate structural characteristics of the network, networks subgroups or single actors (e.g. centrality) to outcome measures like innovativeness or performance (Gulati, 1998). However, studies linking the cooperation partner selection strategies with the evolution of a network are rare. Only a small number of empirical studies have analyzed network formation, dissolution and evolution (Walker, Kogut and Shan, 1997; Orsenigo, Pammolli and Riccaboni, 2001; Gay and Dousset, 2005; Giuliani, 2010; Balland, De Vaan and Boschma, 2012).

This brings me to the following research questions:

- What are the mechanisms and forces that determine the evolution of networks over time?
- What influence do knowledge related factors have in a German automotive innovation network?
- How do firm characteristics affect the propensity to collaborate?
- To what extent do dyadic characteristics influence the preference for cooperation?
- To what extent do endogenous forces drive the evolution of innovation networks?

The analysis is focused on a sample of German automotive firms, taking industry and firm characteristics, such as their collaborative ties and knowledge-bases, into account. I use archival information from a public German database called "Förderkatalog" to learn about interorganizational relations in the German automotive industry based on (subsidized) collaborative research projects. Marsden (1990, p. 444) points out that "only a limited methodological literature exists on archival network data". Accordingly, I also contribute to this literature by exploring the possibilities for applying such data. Furthermore, while the breadth and the depth of firm knowledge-bases have been the subject of previous studies, the structure of the knowledge-base and also the methods of analysis for studying such structures are largely unexplored (Yayavaram and Ahuja, 2008). Therefore, I apply the method of network analysis to investigate the structure of the sample knowledge-base (Saviotti, 2009) which is created out of the patent portfolios of the firms which form the investigated automotive innovation network. I further test a theory derived from Granovetter's (1973) idea about the *strength of week ties,* stipulating that the strength of a tie between two nodes correlates with degree of overlapping ego networks. Moreover, I expect that the current shift in the automotive power train from the internal combustion engine towards e-mobility becomes visible in the centrality measures of respective knowledge elements (approximated by patents).

This brings me to further research questions:

• Do e-mobility related International Patent Classification (IPC) sub-classes become more central in the knowledge-base network over time?
• Does the tie strength among knowledge-base elements (IPC sub-classes) explain the overlap of the elements' ego networks?

To answer these research questions, I use established social network analysis (SNA) techniques and combine them with recent methodological developments in the analysis of network evolution. In particular, I apply the *stochastic actor-based model for network dynamics* (Snijders, 1996; Snijders, 2001; Snijders, 2005). Thereby, I follow the suggestions of the Neo-Schumpeterian school (Hanusch and Pyka, 2007b) and integrate ideas from different disciplines, namely economics, economic geography, management science, sociology, biology and physics (complexity science). Some of the analytical work is done with the help of the network analysis software Ucinet 6 (Borgatti, Everett and Freeman, 2002), but most of it is conducted with software packages that are implemented in the *R* environment. *R* provides a large variety of statistical and graphical analysis techniques. It is an Open Source[3] project and is highly extensible and versatile. Specific packages for network analysis are available. *R* is especially convenient for the analysis of longitudinal network data as it allows for the

---

[3] GNU general public license.

use of predefined functions and loops which reduce manual calculation efforts for larger numbers of networks. Meanwhile, a growing community of R-users develops new packages for specific purposes and shares ideas and codes.[4] Following the philosophy of sharing scientific knowledge, I add my R-scripts for actual analysis to the appendix (C. Appendix). This should allow other researchers (i) to benefit from what has already been thought and make research in the best sense cumulative, and (ii) to facilitate the replication of my results.

The first part of this dissertation is dedicated to theoretical considerations focusing on the complexity of multi-agent systems, the knowledge-based theory of the firm and network structures. The second part continues with an empirical study focusing on the drivers of network evolution in a German automotive network. I test whether knowledge related factors (absorptive capacity, technological distance between actors and the modular character of their knowledge-bases) influence the preference structure with regard to the selection of a cooperation partner. In addition, I test the influence of the geographical distance between actors, their collaborative experience, age, industry experience as well as the preference for being embedded in cohesive triads. The research questions raised in this dissertation are relevant for managers and policy makers alike. First, if co-location improves the chances of becoming embedded in an innovation networks, managers should keep this in mind when they decide about firm location and policy makers have a strong argument for cluster policies. Second, R&D activities can be aligned with network partners that apply a similar knowledge-base leading to small technological distances, or they can be carried out on the basis of diverging knowledge-bases, emphasizing a greater variety of knowledge. For the formation of a network it is important to know if the actors in the industry have a preference for partners with similar or rather dissimilar knowledge-bases. Moreover, a significant relevance of cooperation experience for partner selection suggests that trust and reputation matter, and that there might be a need for special incentives to cooperate with relatively young firms which have not yet had the chance to build trustworthy relationships. In general, the applied model can help to assess if the applied policy achieves its objective in that collaborative ties are formed between the actors that are expected to form the ties, e.g. inclusion of start-ups etc. In the empirical study of the sample knowledge-base, i.e. the aggregated patent portfolio, I investigate, among others, the centrality of technology classes which indicates how important a

---

[4] A good example is the website: http://www.r-bloggers.com.

certain technology is for the industry. This provides guidance for firms when they adjust their research strategy.

This dissertation is divided into ten chapters. Following the introductory chapter, the second chapter illustrates why network embeddedness is a prerequisite for knowledge transfer and social learning. Moreover, I delineate how a Schumpeterian type of competition rewards firms with above-average learning capabilities which allow them shaping the structure of an industry. In the third chapter, the concept of a knowledge-based economy as well as the innovation relatedness of some basic economic models is depicted and discussed. Additionally, this chapter introduces the evolutionary framework and its applicability for the analysis of economic problems. Chapter four is devoted to the theory of the firm which adequately explains collaborative behavior. With regard to an increasing knowledge orientation, the knowledge-based view seems to be an adequate basis for model building. In chapter five, I present methods of social network analysis and important findings from the study of innovation networks with regard to network positions and tie characteristics. Chapter six introduces key ideas of agent-based modeling (ABM) and shows its applicability for the analysis of evolving complex adaptive systems (CAS) and in particular of networks. Moreover, I depict peculiarities of longitudinal network studies. In chapter seven, relevant theories regarding the drivers of network evolution are presented. Network evolution is simultaneously driven by exogenous and endogenous forces. The presented theories constitute the conceptual framework for the development of hypotheses to be tested in chapter 9.10. Obviously, a key element in a knowledge-based view of the firm is the knowledge-base as such. As discussed in chapter eight, a patent portfolio is an appropriate proxy for a knowledge-base and contains additional rich information about inventors and applicants for an innovation economic analysis. In chapter nine, first, I characterize the situation of German automotive suppliers and manufacturers as well as their organization of R&D processes. Moreover, I present the analysis of the automotive sample knowledge-base, focusing on a number of structural characteristics and on technologies which are related to e-mobility. In addition, I illustrate to what extent the overlap of ties is correlated with tie strength. Subsequently, I introduce a model for capturing the drivers of innovation networks evolution for a network which consists of a sample of German automotive firms that are interconnected by publicly funded R&D projects. In a last step, I estimate model parameters and present the results of hypotheses tests. In chapter ten, I discuss a number of important research results as well as possible steps for further avenues of research.

## 2.    Innovation and Industry Dynamics

Schumpeter (1911) puts innovation in the center of his theory of economic development. In the beginning, he linked innovative success purely to the entrepreneurial success of outstanding individuals in an economy. Thirty years later, Schumpeter (1942) – inspired by the development of US industries – identified a significant change in the organization of R&D processes in specialized R&D laboratories of large firms (routinized innovation). And another forty years later, again a significant change had taken place in the organization of R&D. This change refers to the interaction among firms and other innovative actors, such as universities and public research institutes, forming innovation networks. Nevertheless, only since the end of the 1980s has a certain interest in the theoretical explanation of this phenomenon of collective innovation begun to arise in economics and related scientific fields, and the prevailing view of technological knowledge as a quasi public good begun to be challenged. Accordingly, it is not astonishing that we observe differences in firm performance, even in cases where the codified parts of a technology are commonly known. Such differences can (at least partly) be explained with heterogeneous levels of embeddedness in innovation networks (Dosi and Nelson, 2010).

### 2.1    The Case for Network Embeddedness

The notion of an *innovation network* is frequently used in conjunction with agglomeration concepts such as clusters, industrial districts or (regional) innovation systems. The common underlying idea of these concepts suggests that spatial concentration matters for innovation activities on the micro level and for the related economic development from a macroscopic view. Marshall (1920) studied manufacturing firms that were located in the north of England when he brought forward the argument that their success is related to the localization within an industrial district that conveys the – intended or unintended – exchange of ideas: "If one man starts a new idea, it is taken up by others and combined with suggestions of their own; and thus it becomes the source of further new ideas" (Marshall, 1920, p. 271). Within an industrial district, he further argued, knowledge is easily accessible by everyone: "The mysteries of the trade become no mysteries; but are as it were in the air" (Marshall, 1920, p. 271). As sort of a virtuous circle, firms absorb ideas developed by other firms in an industrial district, elaborate on them, learn from them and combine them with their own knowledge, creating in this vein the seed for new ideas which can be grown into even more ideas spreading all over a district. Marshall's discovery provides an argument for spatial clustering of firms. Furthermore, it shows that specialization within an industrial sector exerts the before mentioned externalities that favor the development of the entire spatially concentrated industry.

In contrast to the specialization argument of Marshall (1920), Jacobs (1970) stresses diversity claiming that the exchange of knowledge between industries is more beneficial than exchange within an industry. Diversity enables a spatial cluster to bring new products to the market that are based on a combination of the variety of knowledge which is found in the local economy, thereby increasing the diversity of products which indicates a high innovation output. Recent work on clustering indicates the importance of knowledge exchange with firms that do not belong to the same cluster. Being embedded not only in local clusters but also in global value chains or sources of knowledge is relevant, above all, for firms, regions and countries seeking to catch up with respect to their innovation performance (Bell and Albu, 1999; Giuliani and Bell, 2005). A purely within-cluster R&D orientation creates the risk of getting trapped in a situation of a lock-in as cluster knowledge tends to become gradually more and more uniform over time (Asheim and Isaksen, 2002) following a possibly unfruitful technological trajectory (Grabher, 1993; Cantwell and Iammarino, 2003). The formation of extra-cluster linkages enables a cluster to prevent this lock-in (Asheim and Isaksen, 2002). In addition, Wuyts et al., (2005) find that too close and stable relations are harmful for learning since they reduce the optimal cognitive distance and hence learning opportunities. The integration of less similar external knowledge is a way of increasing the variety of knowledge in a cluster and to counteract the tendency of moving towards uniformity (Ghoshal, 1987; Fleming, 2001).

Technological breakthroughs possibly require a lower degree of cluster or network integration than a potential "optimal" degree which is suggested by transaction cost economics. Uzzi (1997) explains the paradox character of embeddedness: While on the one hand it helps a firm to adapt to the conditions of an ecosystem, on the other hand it hampers adaptation due to decreasing diversity of partners and knowledge-bases. The feeding of R&D processes with external knowledge raises the quantity and variety of knowledge elements which can be recombined with existing knowledge-bases (Fleming, 2001). However, effective knowledge transfer requires some but not too much of cognitive or technological proximity between industries or knowledge-bases in a region. Frenken, Van Oort and Verburg (2007) find that the higher the *related variety* in a region is, the higher the observed regional growth. Boschma and

Iammarino (2009) confirm that inflows of extra-regional knowledge, which is related (but not identical) to the knowledge-base of a region, is an explanatory factor for regional growth. It is important to keep in mind that innovation does not only come from adding new knowledge elements to the knowledge-base but also from adding new links to knowledge elements (see on that point also chapter 7.4 on modularity). The importance of diversity to the consistent creation of novelty is found in a number of industry studies. For instance, Hoang and Rothaermel (2005) find for the biotechnology industry that the higher the observed fluctuation in R&D partnerships, the higher is the performance of the network. Wuyts, Dutta and Stremersch (2004) discover a similar result for the technological diversity among alliance partners. In contrast, Goerzen and Beamish (2005) find for the case of multinational enterprises that diversity in partners diminishes innovative performance. The preference for diversity, or rather conformity, in partner networks will be dealt with in the empirical study of this thesis.

Research not only influences the development of technologies, but it is the discoveries in research that change the organization of the research process as such. For instance, new connections between disciplines may require close cooperation for exploration and further progress. Acknowledging the advantages of interdisciplinary research is one thing, organizing effective interdisciplinary research is, however, another issue. According to Rosenberg (2009, p. 241), "it is unlikely to be successfully planned. Success in the academic world has often failed when administrators have simply decided to form a committee, or program, of researchers from a variety of different disciplines." The rate of success is often higher if there is a prevalent impression within a certain discipline that progress can only be made by integrating solutions stemming from other disciplines (Rosenberg, 2009).

While agglomeration and diversity are enablers for the innovative success of groups of firms, they cannot explain alone interfirm heterogeneity in innovative performance. Recent studies raise serious doubts about the existence of costless knowledge spillovers within a local district challenging the public good character of knowledge (Pyka, Gilbert and Ahrweiler, 2009). Gaining a sound understanding of the interaction between actors is seen as more promising for understanding the advantages of local clustering. Industrial districts are conducive for successful interaction as it is generally easier to talk to geographically co-located neighbors than it is to talk to a remote person. However, it is the development of network structures which explains successful agglomeration (Lechner and Dowling, 1999; Pyka, Gilbert and Ahrweiler, 2009). Even in fields from which we may expect that the relevant knowledge is highly codified, networks as channels and conduits of knowledge transfer play an important role. In fact, "the particular content of the relationships represented by the ties is limited only by a researcher's imagination" (Brass et al., 2004, p. 795). Breschi and Lissoni (2001) find for the mechanical cluster of the Italian region of Brescia that – despite the relatively high degree of knowledge codification – knowledge diffusion is rather limited to a small group of firms. Medda, Piga and Siegel (2006) discover for Italian manufacturing firms that collaborative R&D increases productivity. Also in science-based industries such as biotechnology, chemistry or computer science, it holds that what is publicly known among experts of the same field is only complementary to more tacit and specific knowledge which is essential for the creation of novelties (Dosi, 1988).

A conceptualization of knowledge which implies that knowledge can be acquired like a glass of *cornichons* from a supermarket shelf – but for free – is being replaced by a concept according to which a firm needs to be embedded in a network to absorb knowledge. On the other hand, it would be wrong to assume that a firm's knowledge-base is totally private and secret without any evaporating knowledge elements (Nelson and Winter, 1982). In a nutshell, economically valuable knowledge is not flowing freely in the air but necessitates (costly) efforts to gain access and make use of it. Consequently, the network concept has a different connotation compared to the cluster or innovation system theory. The real value of a cluster, or respectively of an innovation system, is a function of the network ties that are created between the actors (Gulati, 1998). Relations spur innovation processes by providing access to the knowledge-base of other organizations, thereby constituting an element of a firm's organizational capital. For this reason, network embeddedness is a vital asset constituting a firm-specific element of heterogeneity (Granovetter, 1985; Loasby, 2001; Buchmann and Pyka, 2012b). Technological spillovers are not freely available as in the standard models of growth theory but have to be acquired actively by participating in innovation networks. Moreover, Rosenberg (1990) suggests that access to a network of firms is not costless either. Given that a firm seeks to get something out of it, it also has to bring something in. The results of a firm's own research serve in this sense as a ticket to enter a network.

While access to sources of knowledge requires embeddedness, firms also need to be able to absorb the knowledge that is accessed and exchanged through a network structure (Cohen and Levinthal, 1990). Thus, they need to develop a capacity which enables them to recognize valuable knowledge, to understand it and to make something out of it internally by combining it with existing knowledge. Understanding this process is part of the answer which is asked in the title of an article by Rosenberg (1990): "Why do private firms perform basic research (with their own money)?". Investing own money for research would be absurd if knowledge was a quasi-public good, i.e. absorbable without costs "from the air". This would imply that own research results could not be protected and other firms could simply free ride. This "problem" raises the question about the appropriability of inventive outcomes. Moreover, it is often rather unclear – especially in basic research – if efforts will lead to any meaningful innovation that yields benefits in the form of new products or processes. A high level of uncertainty and the risk that the invested money is sunk without generating returns are strong arguments for a firm not to invest in R&D.

## 2.2 Learning and Schumpeterian Competition

Network embeddedness for knowledge acquisition is of particular relevance if we assume a Schumpeterian competition. This type of competition is defined as a "process through which heterogeneous firms compete on the basis of the products and services they offer and get selected with some firms growing, some declining, some going out of business, some new ones always entering on the belief that they can be successful in this competition" (Dosi and Nelson, 2010, p. 96). Competition is driven by innovation, adaptation and imitation. Selection is an endogenous process in which the knowledge absorbing and learning capabilities are important selection criteria in that they create winners and losers in competition.

A firm's abilities to learn, to innovate but also to imitate are regarded as central determinants of industry dynamics in evolutionary economic theories. In competition, two forces are setting the tone, namely idiosyncrasy of firms and market processes which generate profits or losses for firms and are thus selective in terms of survival probability. Learning processes are firm specific but also incorporate a strong collective element (Pyka, 1999). For instance, firms operating in the same industry have similar experiences when they are faced with emerging technological trends. In addition, the search process is a cumulative process: "What the firms can hope to do technologically in the future is narrowly constrained by what it has been capable in doing in the past" (Dosi, 1988, p. 1130). Firms learn from each other through different channels of interaction and from the part of knowledge which was generated within a firm and becomes publicly known (Dosi and Nelson, 2010). This is, however, not sufficient for successfully imitating other firms' inventions. Also, tacit and idiosyncratic knowledge are essential for the development of any kind of technology, be it an imitation or an innovation (Polanyi, 1967; Nelson and Winter, 1982). Such knowledge represents constituent elements of firm heterogeneity, yet they are not totally bound to a single firm but can be overarching such as labor mobility. Diverging learning capabilities, the ability to innovate, the adoption of external innovations, different propensities for investments and adaptation of the organizational structure lead to disparities in firm performance (Dosi and Nelson, 2010). Among the enumerated factors, the ability to innovate seems to be the one which is most unevenly distributed within any given population of firms. A confirming indicator is the distribution of growth rates (which is correlated with innovativeness) which is often highly skewed. This observation can be made across industries and sectors (Bottazzi and Secchi, 2003).

Learning is not only a prerequisite for innovation but also plays a role for imitation. Yang, Phelps and Steensma (2010) demonstrate that the original innovator may benefit from imitations which are often improvements of the original idea. The knowledge which is incorporated in the original invention gets recombined with complementary knowledge of the imitating firm. Thereby, a new pool of external knowledge is established which is related to the knowledge-base of the inventing firm. As a matter of fact, this pool contains knowledge which can be easily understood by the originating firm since it has a high degree of relatedness with the originating knowledge-base. Thus, the original innovator rather easily learns from what the recipient firm has added to the original idea. Learning vicariously from other firms is a type of heuristic search. Organizations learn vicariously by observing the behavior and associated performance outcomes of other organizations and imitate behaviors that seem successful and avoid behaviors that seem unsuccessful (Cyert and March, 1963). By observing the innovative activities and outcomes of other organizations' efforts, a firm can develop a cognitive model of how and why new combinations of knowledge are formed without the need for own experiments. Yang, Phelps and Steensma (2010) illustrate the example of Eastman Kodak. This firm developed in the 1980s an innovative light-emitting diode (OLED), a technology which is applied today in computer and TV screens. In the following decade, more than thirty firms (for example Sony and Xerox among others) benefited from this innovation and allocated own resources to further elaborate on the initial invention. This resulted in a number of additional innovations. Kodak, in turn, learned from these improvements of their original idea and was able to create additional own innovations.

This example also confirms the assertion that knowledge is not purely a public good. In order to make meaningful use of the ideas and developments of other actors, own resources need to be invested. The fact that the transfer of knowledge is not free of costs limits spillovers and at the same time creates incentives for actors to invest in innovative activities. Firms should also be aware of additional learning opportunities which are created when others pick up their ideas and elaborate on them (Yang, Phelps and Steensma, 2010). Also, by introducing the concept of an external knowledge-pool which is related to a firm's original knowledge-base, the concept of spillover becomes highly firm specific in contrast to theoretical considerations which suggest that transfer happens within the same industry (Henderson and Cockburn, 1996) or by applying the same technology (Jaffe, 1986). Thus, innovating firms may even learn from the firms that imitate their products. This positive view of knowledge transfer for the inventor goes beyond a concept which considers knowledge transfer as harmful to the inventor and only beneficial to the imitator, and thereby lowering potential appropriability gains and thus the incentive to invest in R&D projects (Jaffe, 1986; Kogut and Zander, 1992). Learning from imitators and other attempts, such as hiring experts or reverse engineering, qualify as means for extending a firm's knowledge-base, even though they are often not comparably as effective as learning in innovation networks.

## 2.3 Learning Patterns Shape Industry Structures

The establishment of new solutions frequently necessitates that firms reach beyond the frontiers of established knowledge. The pioneering search strategy differs across firms and industries. It is a function of the individual experience of a firm, its internal organizational rules, its product portfolio, and its suppliers and clients (Dosi and Nelson, 2010). And even though the importance of technological leadership is prevalent, it is eventually on the market place where the successful firms are selected from the less successful ones (Nelson and Winter, 1982). The sources for successful innovation are scattered and industry specific. This is, among others, the reason why I focus in the empirical part of this dissertation on one specific industry-network composed out of automotive firms. Industries comprise firms in this approach not because they share ex-anti a common technology, managerial techniques or potential customers, but because common characteristics emerge through the interaction between firms. While formal R&D investments may play a role for developing new goods and services, it is a firm's learning capability that can make the difference in performance (Freeman, 1994). Firms are learning organizations which learn from various sources, including the environment, their rivals as well as from their own successes and failures. Competitive success is thus a function of learning success. In addition, the learning process as such is specific and related to the underlying technological paradigm. Consequently, the evolution of an industry is shaped by the learning patterns (Dosi, 1988).

Learning in networks can be referred to as "social learning" which means that actors learn from others by observing their behavior and above all by directly interacting with them. This type of learning is widespread in nature and is hence expected to be advantageous in the evolutionary process of species. To a large extent, knowledge can only be acquired within a social context and it is often the cheapest way since it reduces efforts and risks which are prone to trial and error learning of own R&D. Yet, there is still the risk that things are learned which are misleading, outdated or not fitting to a firm's knowledge-base. Rendell et al. (2010) organized a simulation tournament on learning strategies with the aim to identify the best performing strategy in competition. In this game, actors could learn or imitate decisions from other actors and the simulated agents could choose between three possible moves in each round, namely innovate, observe and exploit. The innovation strategy consisted of asocial learning which refers to learning through interaction with the environment. Trial and error would be an example for this category. The second strategy, observe, represented the social learning strategy which allows actors to imitate the behavior of other actors by observation or by interaction. The last strategy, exploit, referred to the performance based on an agent's own repertoire. The result of this experiment demonstrates that social learning is a remarkably successful strategy, leading to the highest payoffs even if the costs for asocial learning are not higher than costs for social learning. The winning strategy in the experiment relies almost solely on social learning even though the established view suggests that social learning should be applied occasionally only in order to avoid the absorption of irrelevant or misleading knowledge. It is well known from evolutionary biology that the species which can exist with the lowest level of resource demand has the highest chances to survive. This concept transferred to and tested with the social learning experiments suggests that the dominant strategy is the one which keeps working with the lowest frequency of the very resource intensive asocial learning (Rendell et al., 2010).

## 2.4    Conclusions

If we consider knowledge as a (partly) private good rather than a (pure) public good, then there are learning opportunities or rather necessities for firms which are costly and time consuming. We can conclude that "industrial performance and industrial structures are endogenous to the process of innovation, imitation and competition"

(Dosi, 1988, p. 1157-1158). Learning is related to the exploration of particular fields of (perceived) technological opportunity, to the elaboration of search routines and skill improvement in developing and producing new solutions. Differences in the ability to learn and to innovate are main drivers for asymmetries in industry structure. Firms which come up with successful innovations are able to improve their position in competition.

The diffusion of an innovation and the imitation by competitors should in theory allow all firms to catch up which would result in a convergence process and finally in a balanced industry structure. However, the extent to which a convergence process actually takes place depends on individual firm capabilities limiting the overall tendency for convergence. Often, even a process of increased divergence rather than convergence takes place. The best performing firms forge ahead and force low performing ones to leave the market, a process which increases the aggregate performance of the industry (Dosi, 1988; Rendell et al., 2010). Learning capabilities are not the only technology related feature which influences the industry structure. Also different technological opportunities influence the extent to which the better leaning firms can exploit the potential to improve their performance. Within an environment that offers many opportunities the good learners can distinguish themselves more easily from the bad learning firms which are consequently faced with a high selective pressure. The emerging result is a concentration in industry structure diminishing the chances for firms that are lagging behind to survive (Dosi, 1988).

Static, formal and closed ways of firm organization hamper the process of mutual learning. Instead, open boundaries enable firms to gain access to the external sources of innovation: The networks of research institutes, suppliers, customers and competitors (Powell, 1990). Openness not only concerns firm boundaries but also network boundaries in order to channel new knowledge to the network and increase its variety. Various studies suggest that there is a positive correlation between the intensity and quantity of cooperation in a sector and its R&D intensity, respectively technological progressiveness (cf. Freeman, 1991 and Hagedoorn, 1995). To make use of the learning opportunities which networks provide, firms need to invest in their absorptive capacities and learning capabilities.

# 3.    Methodological Framework

This chapter deals with the fundamental ideas governing the analysis and interpretation of innovation-economic data as well as with the development of a model for explaining innovation network evolution. In particular, the knowledge-based approach and evolutionary concepts in economics are depicted. Knowledge is a peculiar kind of resource and played an important role for the development and performance of economies throughout all times. With rising development levels and with an accelerated technological progress, its importance relative to other types of resources has been even growing over time (Buchmann and Pyka, 2012b). Consequently, the term *knowledge-based economy* became popular among economists as well as among politicians. Knowledge-based economies are "directly based on the production, distribution and use of knowledge" (OECD, 1996, p. 7).

## 3.1    A Knowledge-Based Approach to Economic Development

The knowledge of a firm is its key resource which brings it in the focus of the analysis (Das and Teng, 2000). Accordingly, a firm can be described as a „repository of productive knowledge" (Winter, 1988, p. 171). A key feature of knowledge is its close relation to other firm resources, its specificity as well as its lacking substitutability and the uncertainty in its generation process (Lippman and Rumelt, 1982). The thereof derived heterogeneity between firms has proven to be often stable over time (Peteraf, 1993). Knowledge in the economically relevant sense can be embodied in human beings (as *human capital*) or in goods and services (machines, technology, processes, patents and the like) constituting an organization's knowledge-base. Human capital can be defined as "knowledge, skills, competencies and attributes embodied in individuals which facilitate personal, social and economic well-being" (European Commission, 2003, p. 14). Human capital is often regarded as a strictly economic factor of production, but beyond that, it has a more general meaning for social and cultural life.

Four key drivers of increased knowledge orientation in modern economies have been identified (OECD, 1996):

- Development and widespread use of information and communication technologies (ICT)
- Accelerated technological progress and new scientific knowledge

- Increased global competition (facilitated by reduced trade barriers and declining communication costs)
- Changing demand patterns due to rising income levels

At the Lisbon Summit of the EU's heads of state and government in the year 2000, an overarching strategy was adopted with the clear goal for the EU "to become the most competitive and dynamic knowledge-based economy in the world, capable of sustainable economic growth with more and better jobs and greater social cohesion" (European Commission, 2000, Annex I). EU policies were streamlined, stressing education, human capital and R&D investments. Especially the growing gap to the US in terms of productivity and GDP growth rates should be narrowed and finally completely closed (Sterlacchini and Venturini, 2006).

Lane and Lubatkin (1998) find that in an increasingly knowledge-based competition, firms need to develop a profound understanding of the knowledge-base they own and how it can be applied in a productive way. Moreover, they need to be able to transform knowledge into processes and products, and to manage their knowledge and capabilities like other physical assets (Buchmann and Pyka, 2012b). Firms respond to an increased competitive pressure (Teece, 1992) by forming alliances in which the abilities to learn and exchange knowledge are vital. In other words: "The success of enterprises, as of national economies, is determined by their effectiveness in gathering and using knowledge and technology" (Stevens, 1996, p. 8). Teece and Pisano (1994) as well as Dyer and Nobeoka (2000) refer to continuous organizational learning as a crucial factor for the development of a competitive advantage. The transformation towards more knowledge-based activities is not limited to a few industries but affects economies as a whole. It involves also changes of institutions, habits and cultural behaviors. Hence, we are not only working in a knowledge-based economy, but we are living in an interlinked "knowledge society" (Figure 1). It is the people with their highly developed cognitive capacity, creativity, learning and understanding capabilities which are responsible for success (Rodrigues, 2003).

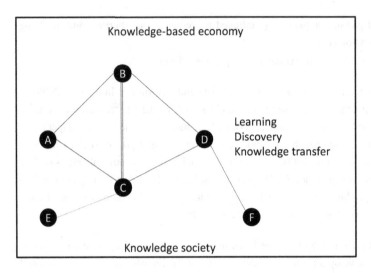

**Figure 1:** Stylized figure of the interlinked knowledge society (Source: own illustration).

## 3.2    The Integration of Knowledge and Technology in Economic Models

The relevance of knowledge for innovation and eventually economic growth suggests that micro-economic knowledge generation, innovation decisions and the interactive character of innovation activities should be explicitly included in economic models. A short overview of the basic principles and characteristics of the neoclassical standard model and the enhancements of the endogenous growth theory shows that they do not qualify for analyzing interactive processes. In fact, the search for the foundations and determinants of economic growth has a long standing history in economic research. There are a number of reasons for this strong interest, not least of which is the versatility of growth as a political goal. High growth rates are expected to increase incomes and general well-being, decrease unemployment rates, fill up the wallet of the minister of finance and rehabilitate social welfare systems, just to name some of them. Economic growth is a collective or aggregate phenomenon which forces us to develop a concept about how to get from the micro to the macro level. Moreover, the study of established models of growth allows for a reflection of their underlying assumptions in the light of previously made suggestions with regard to learning and knowledge transfer processes. Early models of growth concentrated on production functions with two (three) input factors, namely labor and capital (and land). Knowledge and technology were only regarded as external factors that influence production levels.

However, technological progress, the development of knowledge-based industries and the importance of innovation for business success gave rise to the development of new models that integrate knowledge more explicitly in production functions. Particularly challenging is the adequate handling of knowledge investments as they are often characterized by increasing returns, whereas traditional production functions assume decreasing returns to inputs. This latter effect can – in neoclassical concepts – be offset by technological progress (*total factor productivity*) even though neoclassical theory does not provide an endogenous explanation for it. The analysis of historical data shows that decreasing returns are often an exemption than the rule. It is rather new ideas, inventions and innovations which create a counter effect to the relative scarcity assumption embodied in production functions (Dosi, 1988). In the 20$^{th}$ and even more in the 21$^{st}$ century, knowledge became the fastest growing production factor which creates a need for explicitly explaining the role of knowledge endogenously.

In the neoclassical growth model, as it was developed by Solow (1956) and Swan (1956), human capital is neglected as well as heterogeneity of actors. A basic production function takes the form:

$$Y = A f(K, L) \qquad (1)$$

$Y$ is the output, $A$ stands for the total factor productivity (TFP), $K$ for physical capital and $L$ for labor. An increase in $A$ (TFP) is called *Hicks neutral* technological progress. It affects the productivity of capital ($K$) and labor ($L$) likewise. The result changes if we assume for instance only labor augmenting technological progress (*Harrod neutral*). The production function will then be written as:

$$Y = f(K, EL) \qquad (2)$$

$E$ is a labor efficiency index. Technological progress that leads to an increase of $E$ is based on a growth of labor productivity.

Due to the assumed positive but decreasing returns to (physical) capital, long-run economic growth rates cannot be explained by the rate of capital accumulation. With a growing stock of capital the additional output for each additional unit of capital becomes smaller and smaller. Instead, the long-run growth rate is determined by the technological progress and the rate of population growth which are both exogenous variables. Moreover, for the long-run per capita growth rate only technological progress is relevant though it is not further explained in the model. This is rather unsatisfactory and gives little guidance for policy makers to promote per capita growth (Mankiw, 2000).

Mankiw, Romer and Weil (1992) argue that the effect of differences in (the returns of) physical capital becomes a lot smaller when differences in the stock of human capital are taken into account. The extended Solow model encompasses the accumulation of both, physical capital and human capital. The inclusion of human capital lowers the effects of saving and population growth. Accordingly, the basic model is revised and human capital as another variable with positive but decreasing returns becomes introduced in the following form:

$$Y(t) = K(t)^\alpha H(t)^\beta \big(A(t)L(t)\big)^{1-\alpha-\beta} \quad \alpha + \beta < 1 \qquad (3)$$

$H$ represents the human capital stock, $\alpha$ is the production elasticity of physical capital, $\beta$ denotes the production elasticity of human capital and $1$-$\alpha$-$\beta$ is the production elasticity of labor. Consequently, the average level of human capital in an economy affects the level of output in the sense that, for example, an increase in the average level of schooling increases the human capital stock $H$ and hence the level of output $Y$. $L$ and $A$ grow exogenously at rates $n$ and $g$. "$g$ reflects primarily the advancement of knowledge, which is not country-specific" (Mankiw, Romer and Weil, 1992, p. 410). Real income depends on saving rates only (high saving rates lead to higher income). However, also by introducing human capital as a production factor in the suggested way, an increase has only a temporary effect on the growth rate of the economy. The long-run growth path remains determined by exogenous technological progress and by population growth (Canton et al., 2005).

Benhabib and Spiegel (1994) find that human capital growth has an insignificant and negative effect on per capita income growth. They rather postulate an approach which does not take human capital as an additional input in the production function into account, but makes it an endogenous variable. Technological progress (TFP) is modeled as a function of the level of human capital. That is, human capital complements technology in that it helps to adopt and implement it from other nations. Moreover, it influences the capacity of countries to develop innovations which correspond with their production capabilities. Consequently, growth rates may differ across countries in the long run. Furthermore, the level of human capital influences a country's success rate in attracting physical capital.

The endogenous (or new) growth theory was mainly developed to overcome the shortcomings of the original neoclassical model which is based on positive but diminishing returns to production factors and exogenous factors, namely saving rates, population growth and technological progress. This implies that policy decisions have no influence on long-run growth rates. Even if we extend the model and assume that saving rates and capital formation (including human capital) can be influenced, a change in policy would only lead to increased growth in a transition period. The growth path can only be shifted in the short and medium-run. In the long-run output growth is still determined by exogenous population growth and technological progress (Bassanini and Scarpetta, 2001). Most endogenous growth models share the attempt to explain the long-run growth rate as an endogenously driven equilibrium outcome, determined by the behavior of rational actors on markets and by structural characteristics of the economy such as technology and (macroeconomic) policy (Fagerberg, 1994). A simple way to form a concept based on these criteria is to assume that aggregate output varies proportionally with the amount of capital as input to the production process. Consequently, marginal returns to capital are constant, rather than decreasing (Romer, 1986; Lucas, 1988; Lucas, 1990; Romer, 1990):

$$Y = AK \qquad (4)$$

$Y$ is the output, $K$ represents the capital and $A$ is a constant that reflects the technology level. $K$ is assumed to incorporate not only physical capital but also human capital which counterbalances the diminishing returns of physical capital. The capital accumulation process is modeled with the following equation (assuming a positive saving rate $s$):

$$\Delta K = sY - \delta K \qquad (5)$$

Changes in the capital stock ($\Delta K$) equal the investments ($sY$) minus the depreciations ($\delta K$). The output growth rate becomes:

$$\frac{\Delta Y}{Y} = \frac{\Delta K}{K} = sA - \delta \qquad (6)$$

The last two equations indicate that an economy can constantly grow (even without technological progress) as long as $sA > \delta$ (Mankiw, 2000). Note that the per capita growth rate is now determined by a behavioral parameter (among others), namely by the saving rate $s$. However, the saving rate is an average rate of a representative agent without taking heterogeneity, social influence or the relevance of interaction in economic activities into account. Models in equilibrium economics are typically based on the assumption of a representative agent. The framework of analysis requires frictionless interaction whereas economic systems do not show this feature but are characterized by fluctuations and growth or decline (Chen, 2008).

Lucas (1988) stresses human capital as a source for learning and as a means of knowledge transfer via interacting people. "Most of what we know we learn from other people" (Lucas, 1988, p. 38). In this way, firms may benefit from the existing average level of knowledge in an economy. Arrow (1962) and Romer (1986) emphasize in their growth models the role of externalities due to learning (by doing) or knowledge accumulation combined with spillovers from one firm to another in order to counteract the effect of decreasing returns. That is, investments in physical capital increase a firm's knowledge-base and trigger technological progress through "learning by doing".

In these models, new knowledge has the character of a public good and spills over across an economy in the sense of Marshall's (1920) knowledge "in the air" idea without explicitly describing transfer channels and interaction patterns. As a consequence of the public good assumption, there are constant returns on the firm level but increasing returns to scale on an aggregate level. The result is suboptimal investment in knowledge as the private returns are lower than social returns due to benefits from new knowledge which cannot be entirely appropriated. In the respective models, knowledge is regarded as an input to production with an increasing marginal productivity. It is the sum of individual firm's knowledge that can be freely used by other firms and has therefore a positive external effect ($B$) (Barro and Sala-i-Martin, 2004). The respective production function takes the following form:

$$Y = K^{\alpha}L^{1-\alpha}B^{\beta} \qquad (7)$$

The knowledge stock of an economy is regarded as the accumulated sum of individual firm knowledge stocks for instance reflected by the stock of capital:

$$Y = K^{\alpha+\beta}L^{1-\alpha}\text{with } \alpha + \beta = 1 \rightarrow Y = KL^{1-\alpha} \qquad (8)$$

The growth rate becomes:

$$\frac{\Delta Y}{Y} = sL^{1-\alpha} - \delta \qquad (9)$$

In view of recent findings in innovation economics a number of points can be criticized in the presented models:

- The aggregate production function (8) still resembles the "AK model" (4). With elasticity greater than one, the growth rate would continuously increase with capital accumulation. And with elasticity less than one, the growth rate would be exogenous again. Furthermore, the externality depends on a scale effect if it is regarded as the sum of individual stocks of capital. An increase in the number of firms would increase the aggregate capital stock and hence the rate of growth in an economy. Only if we assume that the external effect equals the average capital per worker, scale effects do not matter anymore (OECD, 2003).

- The first generation of endogenous growth models, such as Romer (1990), is characterized by unit elasticity and scale effects. These models suggest that technological progress is a function of the stock of knowledge and of human capital that affects R&D activities. Contrarily, Jones (1995) argues that the prediction of scale effects is inconsistent with empirical evidence. Scale effects imply that by steady increases in R&D, growth rates would augment disproportionately. However, such large effects could hardly be observed even though many countries increased their R&D spendings considerably in the past. The second model generation (e.g. Aghion and Howitt, 1998) does not integrate the scale effects anymore. The fundamental new idea is that an increase in population leads to an increase in R&D activities and human capital, and even more importantly, to new products and industries. Additional R&D is absorbed by new industries. Consequently, the R&D share in each sector remains constant. Hence, not absolute numbers are important but rather the share of R&D investments per industry as well as the share of researches in the working population.

- The argument of learning spillovers is not properly substantiated since learning is in reality time consuming and not effortless (OECD, 2003; Pyka, Gilbert and Ahrweiler, 2009). Moreover, the adoption of innovation and the quest for taking truly advantage of inventions requires considerable changes in a firm's processes and organizational structures. These changes are time consuming which also means that there is a time gap between the initial introduction of, e.g. new machines in the production processes, and potential financial benefits due to superior production techniques (Brynjolfsson and Hitt, 2000).[5]

- Dosi and Nelson (2010) challenge the idea of input factor exchangeability inherent to usual production functions. By using different quantities of inputs the resulting product will most likely not have the same characteristics and cannot be considered the same product. It is consequently not possible to substitute different input factors when relative prices are changing. Instead, new recipes and procedures are required for varying input combinations. A subsequent question is whether small changes in recipes and routines result in small changes in the combination of factor inputs and – linked to this – whether disruptive innovations

---

[5] See for this point also the discussion on the "Solow paradox".

in technologies and thus recipes also considerably change the relative intensities of input factors.

• Furthermore, a problem of the presented models is a lack of necessary complexity which is inherent to innovation and knowledge creation processes. Take for instance possible reverse causality effects. Typical models assume a positive contribution of education to growth rates by increasing knowledge stocks. The causality could yet – at least partly – be the other way round: GDP and productivity growth signify increasing incomes that increase the demand for education if it has a sufficiently high income elasticity. Human capital accumulation receives from this influence an endogenous component (Sianesi and Reenen, 2003). The question arises if human capital accumulation leads to augmented growth rates or if it is rather the other way round. A realistic answer is probably that both directions are working simultaneously in a co-evolutionary manner. But simultaneous and co-evolutionary effects can hardly be tested with the depicted models.

• Chen (2008) criticizes that general equilibrium models having a unique stable equilibrium do not incorporate phenomena we observe in reality such as increasing returns to scale and scope, interactive and strategic behavior in social spaces and product innovation. Moreover, the presented models assume representative agents and neglect nonlinear and collective behavior.

## 3.3    Evolutionary Thinking in Innovation Economics

Evolutionary neo-Schumpeterian models are expected to being better able to replicate and thereby provide plausible explanations for empirically observed phenomena and (stylized) facts in knowledge-driven economies. For instance, a deeper understanding of dynamic selection processes such as the functioning of market processes is required. Therefore, realism of micro mechanisms prevails in such models over the generality of orthodox models. For the exploration of innovation processes, agent-based models are useful as they allow for studying micro mechanisms and emerging patterns in detail, focusing on heterogeneous, bounded rational and interacting agents (Dosi and Nelson, 2010). Evolutionary theory aims at demonstrating how organizational learning and other micro characteristics affect selection and thereby the behavior and characteristic of an aggregate structure. Following Dosi and Nelson (2010) a very basic evolutionary model, taking the selective feature into account, can be formulated as:

$$\Delta s_i = f\big(E_i(t) - \bar{E}(t)\big) \cdot s_i(t) \qquad (10)$$

In this example, $s_i(t)$ denotes the market share of a firm $i$ at time $t$. $E_i(t)$ is an indicator for its competitiveness. Crucial for the market share is the relative fitness of a firm $i$ which is determined by the competiveness of the other firms. Consequently, $\bar{E}(t) = \sum_i E_i(t) \cdot s_i(t)$. Competitiveness is seen as a characteristic which is determined by learning dynamics (see Schumpeterian competition) and by a variety of other factors. The function determining changes in market shares ($\Delta s_i$) has often a nonlinear characteristic.

### 3.3.1 Key Analytical Topics

The evolutionary growth theory (which is also referred to as Neo-Schumpeterian growth theory) seeks to truly endogenize the drivers of economic growth by drawing on ideas originally developed by Schumpeter and further elaborated by scholars such as Nelson and Winter (1982), and Hanusch and Pyka (2007b). A fundamental difference compared to orthodox models is that decisions of heterogeneous economic actors with regard to the creation of innovation are crucial for explaining the growth path. Furthermore, the interplay and feedback loops between individual characteristics and collective features, leading to co-evolutionary processes, are neglected by orthodox theory but explicitly taken into account in the evolutionary theory. That is, orthodox theory typically models firms as a set of decision rules that are linked to external market conditions and internal constraints. Decision rules reflect the assumed objectives in that they specify a number of parameters: the maximization objective (often profits, sometimes more diverse objectives) and the knowledge of the firm (e.g. a production function). The maximization hypothesis is supplemented by the equilibrium paradigm in which the supply-demand equilibrium sets the market prices and determines the behavior of the firms (Nelson and Winter, 1982).

The generality, versatility and relative simplicity of such models needs to be confronted with observations of reality and the researcher's interest. When we read a business newspaper we encounter ambitious entrepreneurs and failing firms, but also successful newcomers withstanding the market pressure and threatening the incumbents, creating new markets before they become themselves chased by motivated innovators with promising ideas and smart solutions. Dynamics is a ubiquitous scheme of capitalism. The explanation of such dynamic change processes in economic systems over time should be a cornerstone of economic theorizing rather than the search for explanations of balanced states. Starting with Schumpeter's (1939) "business cycles", we find plenty of theoretical and empirical evidence for the hypotheses that innovation (technological progress) is a main driving force for industry dynamics, economic development and growth of firms, regions and countries (e.g. Malerba, 2002; Aghion and Griffith, 2005; Perez, 2010). The established neoclassical theory provides yet only limited explanations that can help us to understand the relationship between innovation, dynamics and economic success. Holland (1995, p. 85) stresses this point: "Though it might seem otherwise, market dynamics are not a natural area of study for classical economics". The oversimplified assumption of representative agents and the overly abstracted modeling approach which does not cast light into the black box of the innovation process hinder the search for the micro causes and effects of change processes. As a general line of simplification stylized facts are analyzed in separation rather than considering their interrelated character. A further point of criticism refers to isolated model development which forecloses the integration of knowledge from other scientific fields such as management, organizational science, biology, sociology, physics and history.

Evolutionary economic theory does not deny that firms are profit-seeking organizations; it even recognizes profits as an important firm objective. However, I argue in accordance with Nelson and Winter (1982) that it is doubtful to assume that firms are purely profit-maximizing organizations which can precisely select their strategy from a properly specified (exogenously determined) set of choices. A more realistic assumption is the one made by Simon (1956) and Barnard and Simon (1976) which became known under the notion of *satisficing behavior.* According to this assumption, firms set a minimum level which they want to achieve in terms of profits or other goals rather than firms seek to maximize their goals. Moreover, I refer to the idea of evolutionary processes in economic development determined by Knightian uncertainty (Knight, 1921), to dynamics instead of equilibriums, to heterogeneity of actors and to the concept of bounded rationality which allows us to develop a much more accurate description of firm behavior and to tell a different story of collective innovation and growth. The environments in which firms are embedded are typically highly complex, limiting the extent to which firms may act rational in space and time. Thus, I suggest firm decisions to be myopic, i.e. actors behave only bounded rational with adaptive expectations. These assumptions add complexity because the resulting model is characterized by interaction and adaptive behavior of the actors and absence of a single identifiable optimum solution (Nelson and Winter, 1982).

The optimization and equilibrium approach neglects the level of true uncertainty which is inherent to any real world innovation process (Pyka and Fagiolo, 2005). In fact, uncertainty not only encompasses the lack of knowledge about the costs and benefits of possible alternatives, but it even means that economic actors do not know which alternatives they have (Nelson and Winter, 1982). While risk refers to lacking information about the realization of a known set of alternatives, uncertainty means that the set of possible alternatives is unidentified and the consequences of a particular alternative are unknown. Uncertainty is a particular characteristic for the search of radical innovations which have the power to change technological paradigms thereby triggering market turmoil. The manifestation of a new paradigm goes hand in hand with a reduction of uncertainty (Dosi, 1988). Thus, uncertainty must not be confused with risk. This is clearly described by Keynes (1937, p. 213-214): "By 'uncertain' knowledge, let me explain, I do not mean merely to distinguish what is known for certain from what is only probable. The game of roulette is not subject, in this sense, to uncertainty; nor is the prospect of a Victory bond being drawn. Or, again, the expectation of life is only slightly uncertain. Even the weather is only moderately uncertain. The sense in which I am using the term is that in which the prospect of a European war is uncertain, or the price of copper and the rate of interest twenty years hence, or the obsolescence of a new invention, or the position of private wealth owners in the social system in 1970. About these matters there is no scientific basis on which to form any calculable probability whatever. We simply do not know." In economic systems the degree of uncertainty is naturally high.

### 3.3.2 Evolutionary Models in Biology and Economics

In evolutionary models firms have some though not perfect control over the outcome of their decisions and actions. The existence of bounded rationality prevents firms from optimal adaptation to the same extent as we find adapted organisms (biological systems) in nature. Instead, firms learn and adapt their routines. "A routine is an executable capability for repeated performance in some context that has been learned by an organization in response to selective pressures" (Cohen et al., 1996, p. 683). Routines are firm specific and constitute an element of heterogeneity. The capacity of a firm to conduct R&D and to innovate is to a large extent embodied in the routines that determine a firm's activities. The capabilities of an organization are based on the collection of routines an organization masters (Dosi and Nelson, 2010). Consequently, Dosi and Nelson (2010, p. 81) describe firms as "behavioral entities" which are controlled by "routinized patterns of action". The specific capabilities – it is argued – are the main reason for differences in success rates of firms. The process of routine adaptation is comparable to the recombination and mutation of genes (Nelson and Winter, 1982). However, Nelson (and Teece, 2010) clarifies "[...] I don't have a biological view of economic evolution." While there are clearly similarities, technological, business and economic evolution differ in many respects from biological evolution. For instance, routines differ from genes in that they are frequently subject to changes either by purpose or unintendedly (Nelson and Teece, 2010). By keeping in mind obvious differences between biology and economics, it is still worth studying the communalities of biological and economic systems. Evolutionary biology enriches the analysis of economic phenomena as it provides a broad framework that captures dynamic change processes which we observe in almost all industries. For instance, the development of an industry can be modeled by referring to the concept of selective mating in biology. The mating condition controls the tag of a potential mate if the two fit together. It is then a matter of sufficient resources to produce offspring by crossover, a process which is also conceivable for the case of two firms (Holland, 1995). Hodgson (1995, p. xxi) suggests: "Recognition of the shared problems of complexity in both biology and economics may lead economists to place less faith in methodological individualism and to recognize the legitimacy of levels and units of analysis above the individual." In fact, biological concepts of evolution take different levels of hierarchical selection into consideration. In a nutshell, there is a variety of opportunities for economic theory to draw inspiration from evolutionary biology. However, we cannot transfer all concepts directly but need to decide from case to case if and to what extent a biological framework suits to an economic problem.

### 3.3.3 Evolutionary Biology and Economics

Industry dynamics resemble in essential features processes of natural selection in evolutionary biology which allows us to use the evolutionary framework to model dynamic economic processes. The selective power of markets affects firms but also any kind of technological novelty. Deviating from the direct analogy with biological processes of evolution, selective mechanisms of firms and technologies are not random but only contain random elements. The core determinant of the evolutionary process is the learning capabilities of the actors (cf. chapter 2.2). Therefore, patterns and performance of learning characterize the evolutionary process (Dosi, 1988). Success on markets is a prerequisite to survive and to grow. Firms learn and experience which of their capabilities are related to success and they will seek to retain and renew them persistently (Nelson and Winter, 1982). Individual selection is complemented by group selection. Nelson and Winter (1982) suggest that "evolutionary progress" in a biological sense, that is, adaptation to the changing environment, is what Schumpeter described as the "process of creative destruction".

The idea of transferring evolutionary theories into economic contexts is largely unexploited as inspiration is almost exclusively drawn from Darwin's natural selection postulate. In particular, Spencer's (1874) idea about the "survival of the fittest" together with the views on competition, adaptation and gradual change became a pervasive principle in neoclassical thinking. The strong focus on efficiency has led to a conception of economic evolution similar to the neo-Darwinian view in evolutionary biology. All the change – it is suggested – is the result of steady and progressive changes in efficiency at the level of the individual firm. Going beyond the Darwinian idea, biologists found that species may evolve in a discontinuous form (punctuated equilibrium), whereas, gradual change remains a strong paradigm in economics (Hodgson, 1997).

Whilst learning and the process of discovering new solutions is driven by people such as scientists and engineers that purposely seek to channel and focus their ideas and thoughts, the selection mechanism in biology is arbitrary to the best of biologists' knowledge. This does, however, not totally exclude chance or probabilistic elements from economic models. Indeed, Nelson and Winter (1982) reject a hard distinction between "blind" evolution and "deliberate" goal seeking. They rather suggest that "it is neither difficult nor implausible to develop models of firm behavior that interweave 'blind' and 'deliberate' processes" (Nelson and Winter 1982, p. 11). This is why evolutionary theories for economic change propose to integrate a stochastic element in the models, while they do at the same time not neglect that innovations are purposely created (Dosi and Nelson, 2010).

Evolutionary biology centers on the analysis of group dynamics and opposes the kind of macro analysis which sums up the behavior of single entities or directly assumes a representative agent in the models. In particular, the conceptual distinction between group selection and individual selection is well established, suggesting that both, cooperation and competition, are crucial to understand evolutionary processes in biology as well as industrial dynamics (Van Den Bergh and Gowdy, 2003). A good example which shows how biology can inspire economic analysis is the communality between structural relations of firm interaction and stable triadic relations in a biological context (see chapter 7.7 for a description of triadic structures in networks). Hölldobler and Wilson (1990) describe in their book called "The Ants" a triangular structure which consists of a caterpillar, a fly and an ant (Figure 2). Interaction between the three actors has been observed in the following way: The caterpillar exudes nectar which attracts the fly that lays its eggs on the caterpillar. In this way, the fly becomes a predator through its larva. The third element is the ant which is a predator on the fly. Furthermore, it is attracted by the caterpillar's nectar but has no intention to threaten it. If a larger number of ants surround the caterpillar to degust from its nectar it is somewhat protected from the fly, meaning that it trades resources (the nectar) for protection. If one element of this relationship is removed, the entire triangle falls apart (Holland, 1995). Within an economic context the question can be asked, to what extent can individual actors be removed from a collaborative network without destroying the stability and functioning of the network structure?

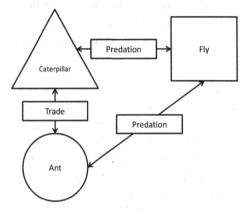

**Figure 2:** Triangle in a biological system (Source: own illustration).

### 3.3.4   Group Selection and Punctuated Equilibrium

Besides the selection of individuals in the Darwinian sense, two more higher-level theories of evolution have the potential for being adopted to describe evolutionary change in an economic system, namely *group selection* and *punctuated equilibrium.* Group selection refers to a selection mechanism that does not solely rely on the fitness of individual actors but assumes that fitness is also a function of a group characteristic. More precisely, it asserts that a group which is voluntarily formed of actors (e.g. an innovation network) showing some degree of altruism supersedes in its performance a formation out of selfish and kin actors. If this assertion was true, it would not only challenge the view of selection at the actor level, but also the *homo oeconomicus* paradigm (Wynne-Edwards, 1991). Group selection is important for economics as it is an even more effective mechanism in human societies than in lower animal populations (Van Den Bergh and Gowdy, 2003).

The altruistic behavior of group members is constrained by the extent of free-riding behavior in the group. Free-riders benefit without bearing costs from the positive characteristics of a group without contributing to the generation or conservation of these characteristics. A problem for the group emerges once the relative share of free-riders exceeds a certain limit as the positive characteristics start to erode. For instance, Oxley and Sampson (2004) find that unintended knowledge transfer hampers firms from cooperating with other actors. It is relatively easy to forge group members to behave in an altruistic rather than self-centered way as long as competition is low and resources are abundant. It becomes, however, less self-evident once the environment changes in a way which makes survival more challenging and requires more adaptation and routine changes (Van Den Bergh and Gowdy, 2003). Giuliani (2005) finds evidence for the relevance of group characteristics, and in particular for the relevance of reputation, in a Chilean wine cluster. By studying the roles of different actors in the cluster it is shown that the so called technological gatekeepers (central actors which are well connected within the cluster and with extra cluster firms) tend to accept unreciprocated knowledge transfer to other firms when positive externalities play a role. That is, increased product quality of single firms due to superior knowledge is likely to improve the reputation of the entire cluster once this knowledge starts to diffuse, an effect which may overcompensate the costs that accrue for the most advanced firms if these firms transfer (voluntarily or involuntarily) knowledge to less sophisticated producers for free. Also, game theoretic models show that cooperation is beneficial in a context that allows for communication among participants and penalization of free-riders (Gintis, 2000). Human beings have tendencies to occasionally cooperate and sometimes purely exploit, thus showing a variety of behavioral patterns that constitute an important element of adaptiveness. Gintis (2000, p. 72) concludes that "seeing human groups as both communities of interacting strategies and (partially) adaptive units deserves to become a major theme in the future".

In short, punctuated equilibrium is a second element of a higher-level theory of selection. It suggests that evolution does not always take place in a gradual manner. Long periods of slow development or even (almost) standstill ("stasis") are interrupted by sudden extensive advances. This concept supplements the theory of gradual change (Eldredge and Gould, 1972) and increases the level of complexity which is necessary to better model real world innovation processes. When Schumpeter elaborated his ideas of disruptive change and creative destruction, he was remarkably ahead of evolutionary theory building since evolution was still considered as slow and only incremental by biologists at that time (Hodgson, 1997). The 1970s stimulated the debate about optimization and gradual changes in standard models with the introduction of the entropy concept drawing on the theory of thermodynamics in physics (e.g. Georgesçu-Roegen, 1971). But only Nelson and Winter (1982) at the beginning of the 1980s manifested the acceptance of nonlinear reasoning in economics.

### 3.3.5 Evolution of Technology

Due to the cumulative character of technological knowledge, technological change may take place for rather long periods in an ordered and only gradual manner, referred to as incremental innovation, thereby stabilizing a technological paradigm (Dosi, 1988). A technological paradigm entails "a 'pattern' of solution of selected techno-economic problems based on highly selected principles derived from the natural sciences, jointly with specific rules aimed to acquire new knowledge and safeguard it, whenever possible, against rapid diffusion to the competitors" (Dosi, 1988, p. 1127). Disruptive changes in technology are related to the appearance of new paradigms. Paradigms encompass both, elements of private and public knowledge (Dosi, 1988). The paradigmatic practice comprises the knowledge about how and why a certain solution works. However, an established practice is in reality not a perfect solution and there is consequently a constant search for better practices that – once established – render old and less effective solutions obsolete. The notion of a paradigm is also linked to the idea of design concepts which describe the properties of products and processes. Moreover, the establishment of a dominant configuration, i.e. a dominant design (Utterback, 1995), goes frequently hand in hand with the enforcement of a paradigm. Paradigms influence the development of technologies in *technological trajectories* that channel progress towards specific directions (Dosi and Nelson, 2010).

The development pattern of new products and processes is shaped by trajectories that limit the degrees of freedom and, thus, the number of possible solutions. Trajectories can facilitate foresight exercises in that they reduce to some extent the degree of uncertainty vis-à-vis future developments in a field: There is only a limited number of new products and processes that can be generated due to the cumulative and paradigmatic characteristics of the knowledge-base. Most trajectories share at least two features: First, one can identify an overarching shift of process technologies towards increased mechanization and automation (Dosi and Nelson, 2010). Second, learning curves and, related to them, falling unit costs, following a power law rule where $p = \alpha \cdot X^\beta$ with $X$ as the cumulated production, $\alpha$ and $\beta$ as constants specific to a certain technology and $p$ representing the unit costs (sometimes also unit labor inputs or an indicator for product performance). Learning effects become manifest on the industry, firm and plant level (Wright, 1936).

*Path-dependency* and the interplay of increasing returns and network externalities may lead to the emergence of a dominant standard. A random event at the beginning of the lifecycle can lead to higher sales of a product and via a feedback mechanism this effect fuels exponentially the use of a specific technology rather than its technological superiority. Also, the cumulativeness of knowledge may trigger snowball effects, i.e. while at the beginning of the lifecycle there is no superiority visible, one technology gets in the initial phase support by different actors (due to various thinkable reasons) and receives consequently additional R&D efforts to further develop this technology and make it more sophisticated. There are indeed two stories of dynamically increasing returns: The first one stresses network externalities. If a large number of users start buying products which are similar or compatible, this process makes it attractive for other users to buy the same standard. ICT-networks, in which all users strongly benefit in having other users have compatible products, are typical examples for this first case. The second story stresses systemic aspects of products. It refers to a product which is accompanied by another complementary product or service that provides the core product with a particular advantage such as computers which gain their real (economic) value from compatible software (David, 1985; Arthur, 1989; David, 2001).

## 3.4 Conclusions

This chapter showed that neoclassical models are based on exogenous innovation to explain economic growth and are hence not suitable to explain innovation processes. To overcome further shortcomings of these models, such as decreasing returns to production factors, the endogenous growth theory was developed stressing the importance of knowledge and innovations. However, these models still suffer from a lack of realism due to stringent assumptions such as rationality of actors, homogeneity of behavior, knowledge spillovers without interaction, equilibrium outcomes and the representative agent assumption. If we follow the suggestions of many empirical studies according to which innovation is a crucial driver of dynamics and collective process, then there are good reasons to focus our analysis more on the process of innovation. A promising way to go is the modeling of innovation processes in a network perspective. This view is based on the interaction of innovative actors thereby forming a complex adaptive system. To understand the functioning of this system, we need to find, for instance, answers to the question which mechanisms account for different attachment patterns. Furthermore, if applied undogmatically, the evolutionary framework is useful to draw inspiration for analyzing dynamic innovation processes.

## 4.    Cooperative Firm Behavior

The innovation process of a firm gets fuelled by internal and external sources. First, scientists and engineers of a firm explore new combinations and exploit new knowledge for the development of new products and processes. This is linked to R&D units which have increasingly become permeable and connected to other departments, such as marketing or directly to customers (Von Hippel, 1988; Chesbrough, 2003). Second, knowledge and new ideas can be accessed externally through links to other actors. Innovation, in this case, is the result of a combination of internal expertise and external stimuli. In knowledge intensive industries firms cannot rely on the internal generation of new knowledge, and the access to external knowledge becomes of vital importance. "Tapping external sources of know-how becomes a must" (Tsang, 2000, p. 225). Alliances are an instrument to bring knowledge and expertise of various firms together for the generation of innovations (Teece, 1992). Note that cooperation can be a highly effective solution for other purposes, too. Examples are: (i) search for stability when firms are faced with environmental uncertainties and threats; (ii) increase of the organizational efficiency; (iii) exercise of power and control over other organizations and (iv) meeting of legal requirements (Oliver, 1990). While in uncertain environments the stability argument is strong, in other situations it is rather the possibility to control competitors which motivates a firm to sign a cooperation agreement (Kogut, 1988).

At first glance, following a cooperative strategy appears like an odd strategy for increasing the output of innovations (and eventually profits) since it may involve the sharing of own knowledge with potential or actual competitors. In particular, firms which operate at the technological frontier endowed with strong commercial skills run the risk of losing valuable knowledge and of strengthening their competitors (Kitching and Blackburn, 1999). A go-it-alone strategy for innovation seems to be more advantageous for them. However, based on a survey of the respective literature, Pittaway et al. (2004) list a comprehensive number of motives for network participation, a pattern which frequently emerges out of bi- and multilateral cooperation (Pyka, Gilbert and Ahrweiler, 2009): (i) risk sharing; (ii) obtaining access to new markets and technologies; (iii) speeding up time to market; (iv) pooling of complementary skills; (v) safeguarding property rights when perfect contracts are not possible; (vi) acting as a key vehicle for obtaining access to external knowledge. Most of these motives are related to changes in innovation processes in knowledge-based economies. That is, firms often cooperate as a reaction to increased complexity in knowledge generation and diffusion processes and to cope with technological uncertainty. Hagedoorn (1993) reports a number of specific motives for network participation that explain why firms benefit from cooperative structures in their R&D processes. In different industries three reasons are important: (i) technological complementarities, (ii) shortening of the innovation time and (iii) market access as well as influence on market structures. To conclude, firms seek to better manage complexity and technological uncertainty by means of cooperation.

## 4.1 Collaboration Characteristics

The literature on interorganizational cooperation lists numerous types of cooperation. Prevalent types are joint ventures, equity alliances, joint production, joint marketing, supplier partnership ("one face to the customer"), distribution agreements and licensing agreements (Gates, 1993). A crude way to classify alliances is to differentiate between equity sharing and non-equity forms of cooperation. When talking about cooperation in this dissertation, I do essentially mean cooperation between firms in joint research projects on a contractual basis (Mowery, Oxley and Silverman, 1996), encompassing no exchange of equity but an exchange of (intangible) resources. Exchange of resources is the result of regular interaction over a certain period in time which widens firm boundaries by partly mutual resource integration.

Winter (1988), by reflecting on the concept of firm boundaries, asserts that the present state of the art is characterized by incoherence and contradictions. Teece (1986) argues that the boundaries of the firm can be found where the scope of a firm's core competences ends. In line with the general reasoning of this dissertation, boundaries are considered as rather fuzzy structures that are permeable for resource inflows and outflows. They function as interfaces for the interaction with external organizations. The challenge for a firm is hence not to build *Chinese walls* in order to separate the inside world from the outside world or to protect internal knowledge from transfer to other organizations. The concern is rather to build appropriate ties and to find ways to share the boundaries with other organizations that allow for an effective knowledge exchange.

## 4.2 Theories of Network Formation

In order to analyze, explain and model (collaborative) firm behavior, a consistent theory of the firm has to be at hand. By taking the before mentioned challenges and motives into account, it needs to theorize the nature of the firm, its operational motives, strategic objectives as well as its boundaries. In neoclassical economics (Solow, 1956; Swan, 1956) firms are represented by a more or less complex production function reflecting the efficiency of the applied production technology. Contrarily, the resource-based view regards firms as heterogeneous entities that are characterized by the resources they own and apply (Penrose, 1959). Whilst the process perspective of a production function approach is akin to the resource based-view of the firm, it is not the production function itself which represents the firm, but the firm is represented by a bundle of unique resources and it rather develops, improves and adjusts its production function (Penrose, 1959; Rumelt, 1997). Marshall (1920) already contributed to the theory of the firm by suggesting that each firm possesses a unique set of relations with other actors. Such relations spur innovation processes by providing access to the knowledge-base of other organizations thereby constituting an element of a firm's organizational capital. Relationships are vital assets for the survival of a firm and likewise a particular firm specific resource constituting an element of heterogeneity (Granovetter, 1985; Loasby, 2001). In the following subchapters I briefly delineate four frequently used concepts of the firm, namely the production function approach, the transaction-cost approach, the resource-based approach and finally the knowledge-based approach.

### 4.2.1  Profit Maximizing and Cost Minimizing Concepts

In the production function approach the firm is seen as a functional relationship between inputs and outputs of production. This production function approach constitutes the neoclassical workhorse of analysis. Accordingly, the questions to be answered are those on the optimality in the allocation of production factors and the respective incentives of firm behavior (Pyka, 2002). With regard to industrial innovation processes, since the early 1980s a branch of literature (new industrial economics) also analyzes the conditions and incentives of firms to engage in R&D cooperation by drawing on a game-theoretic framework (e.g. d'Aspremont and Jacquemin, 1988). Based on a Nash-Cournot type of duopoly model, the outcome of cooperative and non-cooperative R&D is analyzed. Different degrees of spillovers are taken into consideration ranging from no spillovers at all ($\beta = 0$) to complete spillover of the generated knowledge ($\beta = 1$). The maximum profit becomes a function of marginal costs $c$ but also of the R&D intensity $x$ and the degree of spillovers $\beta$ ($A$ is the intercept and $B$ the steepness of the demand function):

$$\pi_i^c = \frac{(A - c + x_1(2 - \beta) + x_2(2\beta - 1))^2}{9B} - \frac{x_1^2}{2} \qquad (11)$$

According to this model, cooperation is beneficial in cases when the appropriability conditions for R&D investments are weak and, thus, the rate of knowledge spillovers among different actors is high ($\beta > 0.5$). Firms cooperate in order to benefit from the other firm's knowledge and to compensate for the diffusion of own knowledge. The higher $\beta$, the more each firm spends on R&D. However, from a welfare perspective, a situation of cooperation with a low degree of spillovers ($\beta < 0.5$) can be disadvantageous since firms may reduce R&D to reduce competition, leading to higher prices and to a smaller consumer surplus.

The second approach can be traced back to Coase (1937) and does not focus on immediate production processes but on transaction costs of economic activities. If we imagine an economy without firms, it would simply consist of isolated labor-sharing individuals connected by coordinating markets. Only the bundling and organization of sets of activities within a firm gives an industry its specific structure (e.g. small and medium-sized firms, large enterprises etc.). However, not only the existence of firms as such, but also their embeddedness in networks, constitutes a decisive feature of the economic structure. For Coase and his followers the main reason for the existence of firms is the costs that accrue from market transactions. Accordingly, firms come into being because the costs of coordinating transactions via markets are higher than the costs of a hierarchical organization within a firm. Essentially, it boils down to incentives for cost savings.

These considerations are transferred to networks for instance by Williamson (1975). In this perspective, networks are an intermediate form of coordination between the dichotomy of hierarchy and markets. With the introduction of uncertainty and specificity of resources within an environment of bounded rationality and opportunistic behavior, networks are considered as a hybrid structure, balancing the costs for controlling an organization and the costs for acting on markets. In order to optimize the organizational structure, a firm has to take the transaction frequency and the importance of asset specificity into account: For intermediate degrees of asset specificity and intermediate levels of uncertainty innovation networks are considered the best organizational solution. With regard to market transactions that involve R&D, costs accrue due to (i) unclear contracts as research is largely uncertain in its outcome; (ii) inability to protect proprietary research results; (iii) risk of getting dependent on research suppliers which unilaterally benefit from a cooperation and (iv) monitoring costs (Williamson, 1975; Teece, 1992).

We may conclude from this approach that firms preferably employ their internal knowledge for research and development activities targeting new products and processes (and other kinds of innovations). External R&D is only used for non-critical activities that do not involve the risk of losing important knowledge. However, in knowledge-intensive industries, it is often less the internal R&D unit that makes all the difference in innovative performance, but it is rather the ability to learn from external actors which leads to a comparative advantage (Powell, Koput and Smith-Doerr, 1996). The crucial questions for the formulation of the innovation strategy is hence less of the kind of to make or to buy; rather, in fields of high complexity where the comparative advantage is bound to innovative performance, the focal point of innovation is put on learning in networks of innovation (Powell and Brantley, 1992; Pyka, 1999).

The above presented two approaches to explain the existence of innovation networks are increasingly criticized. The main critical point is the strong focus cost and profit considerations, that is, among a set of alternatives the less costly or most profitable alternative will be chosen. This assumption contradicts with the concept of uncertainty (Knight, 1921). Moreover, the creative potential of innovation networks, bringing together complementary knowledge and technologies, is not considered. In addition, Powell, Koput and Smith-Doerr (1996) find for the case of biotech firms, representing a science-based industry (Pavitt, 1984), that growing and ageing firms do not reduce the number of collaborations they are involved in. This contrasts with the transaction-cost approach which suggests that firms increasingly use internal capabilities that are quasi free of transaction costs. Transaction cost economics can be characterized as a static cost trade-off analysis. The limitation on cost analysis does not take into consideration the induced possibilities for organizations to absorb knowledge in networks and the importance of social embeddedness in general. Moreover, the concept of transaction costs is vaguely defined and hard to measure, and hence unsuitable to provide guidance in decision making processes (Chen, 2008). From the perspective of a firm, there are strategic reasons other than cost considerations, such as knowledge sourcing, when they enter an alliance (Kogut, 1988; Powell, Koput and Smith-Doerr, 1996). Chen (2008) criticizes typical models which resemble in important aspects a perpetual motion machine, i.e. there are internal firm transactions which are free of costs. Any information collection or transmission requires some form of energy input and produces frictions, and entropy production increases in biological and social evolution.

**4.2.2 A Resource-Based View of the Firm**

The third strand of literature, the resource-based approach, differs sharply from the previous two approaches (Kogut and Zander, 1993; Pyka, 2002). The resource-based view and its particular case, the knowledge-based view of the firm, serve as conceptual tools to shed light on the production function and technological progress, and help to identify its components and interaction patterns (Spender, 1996). Both approaches redefine the firm in ways that move us beyond a mere collection of rational individuals. Early contributors to this theory are Marshall (1920) who recognizes in particular knowledge as the decisive resource in production processes and Penrose (1959) who theorizes a firm by its bundle of resources. Resources are defined as "those (tangible and intangible) assets which are tied semi-permanently to the firm" (Wernerfelt, 1984, p. 172). Tangible assets are typically those assets which enter the firm from the external environment. In contrast, intangible assets such as knowledge are internally created and/or in their applicability highly firm specific. Constraints of a firm which hamper it from growing and enlarging its production facilities are linked to imperfections of the organizational knowledge rather than to external factors. Over time, firms are able to learn and adapt (Loasby, 2001).

Firms seek to increase their value by combining resources according to the most beneficial recipe. The Neo-Schumpeterian scholars show that firms cannot be considered as atomistic entities in perfect markets. Firms constantly adapt the combination of their resources to changing environments and newly emerging paradigms. In the advent of new paradigms, incumbent firms have a hard time to adapt and new firms get chances to enter formerly closed markets (Hanusch and Pyka, 2007b). Each firm is thus as a unique actor, i.e. the combination of resources it applies is unique. Even when formally applying the same combinations, the resulting production process may differ due to a tacit component. Tacit knowledge is of utmost importance in high-tech industries (Jones, Hesterly and Borgatti, 1997). What leads to technological progress on the aggregate level, is the process of imitation and diffusion of existing best practices, the constant striving of firms to improve existing practices, diminishing inferior technologies and fluctuating shares of new and incumbent techniques (Dosi and Nelson, 2010).

If resources cannot be transferred on markets, cooperation is a possibility to anyhow enable a transfer between firms (Figure 3). For example, reputation can be transferred to a strategic alliance formed by two or more firms. This holds for the tacit knowledge of firms, too. The transfer requires trust and personal interaction, and often complementary resources are required in conjunction. Trust between cooperating firms is a key ingredient for the recipe of successful knowledge exchange and it activates a teaching firm to actually understand which knowledge the learning firm seeks to acquire (Johnson et al., 1996). Trust moderates the partner's behavior, reduces the risk of misbehavior (Gulati, 1995a) and regulates the magnitude and efficiency of knowledge transfer processes (Kogut, 1988; Johnson et al., 1996). Reciprocal knowledge exchange and synergistic creativity in innovation networks does not work without trust among the network participants (Almeida and Kogut, 1999). Trust facilitates knowledge exchange and learning in cooperative projects. The concept of interorganizational trust goes beyond individual relationships and develops into administrative routines, norms and values (Dodgson, 1993). Firms may work together in one project while they are rivals in another one. Organizing business in this way requires skills and methods that allow firms to change regularly their partners without destroying the common basis for later cooperation (Powell et al., 2005). Still, there are in practice considerable concerns that the open exchange of knowledge might strengthen competitors. Dilk et al. (2008) find for the automotive industry that only about 12 % of all innovation networks are horizontal. The necessary level of trust can emerge from the cumulative experience of past interactions (Ring and Van de Ven, 1992). This holds even more for informal networks, which play a decisive role in speeding up knowledge diffusion. Firms that follow a strategy of isolation and secrecy, and do not actively seek to exchange knowledge in cooperative structures limit their possibilities to get access to external knowledge in the future since they lose the competences required for participating in cooperative projects (Shaw, 1993).

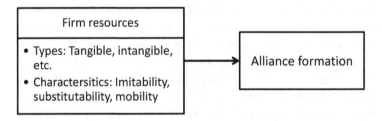

**Figure 3:** Firm resources and their influence on alliances (Source: own illustration).

A number of resource classification schemes are described in the respective literature on the resource-based view. As shown in Figure 3, a basic differentiation is the one between tangible and intangible resources (Grant, 1991). Reed and DeFillippi (1990) assert that the focus on physical assets alone will not be sufficient to attain a high performance level in the long run. Other authors suggest a more fine grained differentiation. For instance, Hofer and Schendel (1978) classify financial, physical, managerial, human, organizational and technological resources. Miller and Shamsie

(1996) distinguish property-based resources (PBR) from knowledge-based resources (KBR) (Table 1). In this concept, the PBRs encompass all assets that are legally owned by the firm. This applies to physical resources and financial resources but also to patents, copyrights etc. KBRs concern the know-how and skills of a firm. The notion of knowledge also encompasses technological and managerial systems which are not protected by legal regimes (Hall, 1992).

**Table 1:** Resource types.

| | Resource Types | |
|---|---|---|
| Resource Characteristics | Property-Based Resources | Knowledge-Based Resources |
| Imperfect Mobility | Human resources | Organizational resources (e.g., culture) Technological and managerial resources |
| Imperfect Imitability | Patents, contracts, copyrights, trademarks, registered design | |
| Imperfect Substitutability | Physical resources | |

Source: own illustration based on Miller and Shamsie (1996).

Resources are characterized in terms of their mobility, imitability and substitutability (Lippman and Rumelt, 1982). The tacitness of knowledge makes knowledge-based resources hard to transfer and to imitate. Organizational resources such as firm culture, firm reputation or the learning and absorptive capacity have deep roots in a firm and are interdependent and interwoven in a way that they can hardly or not at all be retrieved from their original context (Dierickx and Cool, 1989). In addition, knowledge-based resources are inherently uncertain in the development process and difficult to substitute if better technologies are not available. In contrast, property-based resources are more mobile, substitutable and tradable on markets. Grant (1996) sees the main contribution of the firm in the integration of knowledge – which resides within the employees – with their coordination capabilities. Accordingly, the focus is less on knowledge creation but more on knowledge application. In this perspective, the existence of firms is explained by the framework they create for the integration of individual specialized knowledge. This individualist view contrasts with a view which is more focused on knowledge generation and acquisition building up a strong organizational knowledge-base.

In a nutshell, I model firms as knowledge-based organizations acting as "repositories of productive knowledge" (Winter, 1988, p. 169) that are able to adapt, learn and exchange knowledge among each other by interaction. The value of a specific knowledge element is firm specific. While in the example in Figure 4 the value of a knowledge unit $D$ is rather small (small circle) for the firms $j$ and $k$, it may have a much larger value (large circle) for firm *i*. For the analysis of innovation networks and underlying knowledge exchange processes within the scope of this dissertation, I operationalize the knowledge-base by the approximation of its patent portfolio (cf. chapter 8). Imperfect mobility, imperfect imitability and imperfect substitutability of knowledge hamper firms from acquiring required knowledge on markets. An alternative transfer channel is consequently required. Cooperation and subsequently the emergence of networks open such channels and provide incentives to voluntarily participate in transfer processes.

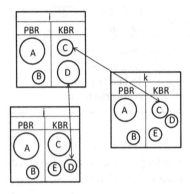

**Figure 4:** Exchange and diverging relevance of knowledge (Source: own illustration).

### 4.2.3   Towards a Knowledge-Based Theory of Interfirm Cooperation

I state in chapter three that knowledge has become a resource of particular importance in modern economies. Consequently, I am looking for a model which adequately captures knowledge generation and exchange processes. In the previous subchapter, I delineate how knowledge can be operationalized within the framework of the resource-based view of the firm. Accordingly, the cooperative behavior of firms is influenced by the scale and scope of firms' knowledge-bases. This allows us to generically develop complex firm aggregates like networks. In this context, four lines of knowledge-based reasoning for network formation are identified: First, network members benefit from knowledge, outsiders do not get access to. It is implicitly assumed that the trading of knowledge on markets is impossible (Eisenhardt and Schoonhoven, 1996; Das and Teng, 2000; Gulati, Nohria and Zaheer, 2000; Vonortas, 2009). Additionally, cooperation allows for the improvement of knowledge-bases by mutual learning. Hence, rather than building vertically integrated organizations, firms prefer intertwined network organizations that provide them with opportunities to learn from each other and exchange a large variety of knowledge. Acquired awareness for the importance of mutual learning opportunities promotes eagerness to collaborate (Powell, Koput and Smith-Doerr, 1996). Second, network actors may influence knowledge flows which provide them with a certain degree of power and influence (Pfeffer, 1978). Third, firms not only cooperate to compensate for a lack of own resources (knowledge) but also to explore and exploit their own knowledge-bases (Powell, Koput and Smith-Doerr, 1996). Kogut (1988) illustrates the case of a firm which cannot make use of a resource it owns at a certain moment in time and which it wants to keep internally "stored" for potential later usage. This might be research personnel which is not used to capacity. A firm may now search for other firms which have different resources available, such as physical resources, that may be meaningful combined with the own research personnel. Fourth, Nelson and Winter (1982) suggest that firms run the risk of degrading their knowledge-base if they remain too long isolated from external stimuli which could reactivate unused elements of their knowledge-base. A firm has to decide whether it wants to permanently externalize the unused resource by a merger or acquisition, or if it rather wants to collaborate for a certain period of time in order to revitalize old capabilities by actually applying them.

In highly competitive environments that are characterized by high velocities of innovation, knowledge intense products and fast-pace market entry strategies, collaborative innovation by transferring knowledge and learning from each other is key to success. Especially in industries which are characterized by a high level of interrelatedness and complexity, finding a complementary partner is a real advantage in competition (Hagedoorn, 1993). In such environments the sources of relevant knowledge are typically dispersed and controlled by a larger number of different firms. Furthermore, the required resources can typically not be separated from other resources a firm owns. Firms collaborate within such constellations to improve their innovation output and speed, and to reduce uncertainty (Ramanathan, Seth and Thomas, 1997). In less competitive environments where the crucial resources are rather concentrated in a single firm, the propensity to cooperate is often less strongly expressed (Eisenhardt and Schoonhoven, 1996).

Following Burt (1995), knowledge benefits are characterized by three distinct features, namely access, timing and referrals. As described above, the first feature is related to the provision of knowledge from partners in the network as well as from potential future partners. Because timing matters a lot in innovation processes, firms benefit also from an accelerated knowledge transfer which is given in network structures (Cowan and Jonard, 2004). Finally, valuable knowledge and information required by a network actor may not be found in the intermediate neighborhood, i.e. from actors a firm is directly connected with. Instead, the specific knowledge might flow through the dispersed channels in the network coming from more remote actors. In an accelerated and complex knowledge generation and diffusion process, therefore, participating in an innovation network supports learning and updating of knowledge-bases of the actors that are members of an innovation network. Besides direct learning opportunities, networks are means to reduce uncertainty about the usefulness of things that can be learned (Galaskiewicz, 1985; Gulati and Gargiulo, 1999).

Ties within innovation networks not only reflect formal contracts but also informal relationships (Hanson and Krackhardt, 1993; Pyka, 1997). Moreover, personal relations between two or more representatives of involved organizations are an important success factor for cooperation (Ring and Van de Ven, 1992; Doz, 1996) since they facilitate the transfer of information (Von Hippel, 1987) and the formation of trustworthy relationships (Gulati, 1995a). Freeman (1991, p. 500) finds in a study: "Although rarely measured systematically, informal networks appeared to be the most important. Multiple sources of information and pluralistic patterns of collaboration were the rule rather than the exception." Dahl and Pedersen (2004) find for the case of a cluster of wireless communication firms in Northern Denmark that informal contacts considerably fuel knowledge diffusion processes. Owen-Smith and Powell (2004) analyze knowledge networks in the Boston biotechnology community and thereby confirm that informal relations between the actors foster knowledge exchange between agglomerated firms. A particular type of an informal network is observed by Von Hippel (1987) as informal knowledge exchange among scientists and engineers working for different and even competing firms. "Informal know-how trading is the extensive exchange of proprietary know-how in informal networks of engineers in rival (and non-rival) firms" (Von Hippel, 1987, p. 291). This exchange is based on trust and personal contacts which are systematically developed as a function of personal judgments of the usefulness and value of the knowledge to be received or to be transferred. Other examples are friendship networks among managers which are used by managers whose firms are in "troubled water" (McDonald and Westphal, 2003). Even though the exchange does not undergo a formal evaluation process, the decision of an individual engineer to trade know-how is reasonable. The quality of advice can be immediately tested by applying it, while the sender can test the quality of expertise of the receiver by evaluating the sophistication of the expressed demand.

Innovation networks have a special meaning for small firms. For example, in the pharmaceutical industry small firms frequently cooperate with the big players not only to benefit from financial resources but also from marketing, juridical and operations know-how, i.e. they gain access to relatively immobile resources. In return, large firms gain the knowledge which they need to develop new drugs and regularly fill up their product pipelines (Müller, 2005). By studying the innovative performance of start-ups in the biotechnology industry Walker, Kogut and Shan (1997) find a positive relationship between the number of collaborative ties a firm formed and its innovation output. The exchange of tacit knowledge involves high transaction costs as it demands close interaction and synchronization of knowledge-bases. Small firms are particularly affected by the problems that are related to the tacitness of knowledge: First, smaller firms often sell highly specialized products for niche markets incorporating a high share of tacit knowledge. Second, work flows tend to be less formal and explicitly written down (Boschma, Eriksson and Lindgren, 2009). Network strategies are important for start-up firms as an entry strategy. Innovation networks provide access to knowledge and other resources which increase the chances to survive (Ostgaard and S. Birley, 1996). Entering established innovation networks opens up such possibilities for young and small firms which are hardly conceivable outside the network (Rothwell and M. Dodgson, 1991).

Miller and Shamsie (1996) state that the protection of knowledge-based resources from unintended diffusion is much more difficult than the protection of property-based resources due to the lack of the applicability of legal rights. Cooperation partners are consequently concerned with the threat of losing their knowledge to a partner (Hamel, 1991). In fact, many studies neglect the stimulating aspects of knowledge transfers (Kogut and Zander, 1992) with the exemption of creating an (unintended) industry standard (Spencer, 2003). However, also the knowledge which is embodied in property-based resources, such as patents, is not fully safeguarded. There are ways which enable firms to disrespect legal boundaries and to exploit other firms' patents, for instance by "inventing around". On the other hand, access to and transfer of knowledge as well as its application is not free of costs either. It requires absorptive capacity and internal R&D in order to recognize and make use of it. Furthermore, even if the unintended diffusion of private knowledge in networks is costly, for instance due to an increase in competition, imitation creates new learning opportunities. The recipient firm will recombine the absorbed knowledge with own knowledge thereby creating a new solution which creates in turn new learning possibilities for the originating firm whose knowledge-base is related to the recombined knowledge-base of the imitator (Yang, Phelps and Steensma, 2010).

## 4.3    Conclusions

Concepts focusing on costs and profits in combination with assumptions of rational behavior and knowledge-spillover can hardly explain the increasing number of collaborative research projects. Resource- and in particular knowledge-based concepts focus on the singularity of firms and their internal characteristics. Resources have a strategic importance for the firm and hence, they need to find ways to gain access to vital resources they do not possess themselves. Varying resource endowments stresses heterogeneity. In contrast, concepts which focus predominantly on external factors, such as the competitive environment, are often too much based on the assumption of firm homogeneity (Dierickx and Cool, 1989; Barney, 1991). Heterogeneity is a persistent property opening the door for the analysis and development of distinctive firm strategies by which a firm can gain an advantage in competition. According to Barney (1991, p. 102), "a firm is said to have a competitive advantage when it implements a value creating strategy not simultaneously implemented by any current or potential competitor". Thus, a superior strategy is based on the combination of resources which creates more value than other strategies. However, the development of a successful strategy is not a straightforward task since it is not obvious which combination of resources eventually leads to a competitive advantage. It is related to the tacitness, complexity and specificity of resources (Reed and DeFillippi, 1990). This problem, also known as *causal ambiguity* (Lippman and Rumelt, 1982), makes it for external persons particularly difficult to identify successful resource combinations. It exacerbates the search for resource substitutes and impedes the imitation of strategies.

# 5.  Concepts of Descriptive Network Analysis

Differing positions in a network are linked to operational opportunities or limitations, especially with regard to access to knowledge. It is not only advantageous to be a member of a network but it is also important to find the right place in the network, most of the time preferably a central position that allows for absorbing the latest developments. Furthermore, evolutionary dynamics is not a phenomenon which is limited to firms or technologies, but also networks change their structure over time. Consequently, the position of actors in the network changes. Also, the number of network members changes as new firms enter and incumbents leave a network. The topology of a network reflects the channels through which knowledge can be exchanged. To study the structure of a specific network, *social network analysis* (SNA) becomes applied. The basic methodology as well as a number of important findings which are drawn from the application of SNA to innovation networks will be presented in this chapter. However, to understand network evolution, studying a snapshot structure is only a first step. More information about the causes and consequences of dynamic interaction is required (Barabasi, 2007) (cf. chapter 6). For instance, many real-world network structures exhibit scale-free structures in which a relatively central node that has already many links gains further links more rapidly compared to a weakly connected node. The result is a positive feedback mechanism which favors the ones which are already well connected. More complex attachment mechanisms are delineated in chapter 7.

## 5.1  Features of Natural Networks

Social network analysis has become a widely used tool for the analysis of interaction processes in many scientific disciplines. In the first chapter of this dissertation, I referred with the cooking example already to the biological context in which network analysis is indeed a prominent method of analysis. For instance, the functioning of a biological cell can be described as a metabolic network with enzymes and substrates representing the nodes and chemical interactions representing the ties. Another biological network is our brain which consists of nerve cells connected by axons (Barabasi, 2007). Barabási and Oltvai (2004) find that structural characteristics of cellular networks resemble other complex systems such as the internet or society. They conclude that the laws governing most natural complex systems must be similar. Suda, Itao and Matsuo (2010) study transport networks in social and biological systems with a special interest in the robustness of network performance. Relevant aspects in this context are costs, transport efficiency and fault tolerance. The basic hypothesis is that biological networks have been shaped by the pressure of evolutionary selection which should have given them a high performance level. To test this hypothesis, the slime mold *Physarum polycephalum* is compared with the Tokyo rail system. The main challenge in transportation is to strike a balance between the costs of generating an efficient network and failure resilience. A "biologically inspired model for adaptive network development", which was derived from the study of the slime mold, was in experiments able to produce solutions which were performing at least as well as real-world infrastructure networks. A widespread feature in natural networks is the scale-free architecture (Barabasi and Albert, 1999; Barabasi and Albert, 2002; Barabasi, 2007) contrasting the random structure of early network models such as Erdős and Rényi (1960). Scale-free indicates that there is no typical node in the network, that is, a node which could be regarded as typical for a certain network. The emergence of scale-free networks can be explained by a preferential attachment mechanism (Müller, Buchmann and Kudic, 2013). In presence of a preferential attachment mechanism, the likelihood $P$ for a new node to connect to an incumbent node with $k$ links is proportional to $k$. $P(k) = \frac{k}{\sum_i k_i}$ with the sum encompassing all network members. In contrast, random networks have nodes with degree centrality measures deviating only slightly from the average degree.

## 5.2    Collecting Network Data

The first step to social network analysis is to collect data of nodes and ties. Network data about interpersonal or interorganizational networks is typically collected through surveys, questionnaires, from archived documents, observations or electronic traces (e.g. E-Mail, Facebook) (Newman, 2003). The reconstruction of networks based on interviews or questionnaires often suffers from inadequate sample sizes or from a subjectivity bias. An alternative method for capturing communication networks is the analysis of communication records. Thereby, the assumption is made that a tie between two persons comes into being if an email is sent from one person to another (Newman, 2003). With increasing computational power, scientific fields such as biology and physics have been reframed by the capacity to collect and analyze large data sets. Deviating from this development, Lazer et al. (2009) observe that the application of large data sets within computational social science has been less pushed and "to date, research on human interactions has relied mainly on one-time, self-reported data on relationships" (Lazer et al., 2009, p. 722).

Collecting network data to analyze social networks in the widest sense can be quite cumbersome as a number of preconditions have to be fulfilled for a sound analysis: First, some methods of analysis require complete network data which is often difficult to guarantee (Freeman, 1979; Winship and Mandel, 1983). Second, some sort of network boundary has to be specified. This is often done by (i) applying geographic boundaries; (ii) by formalized membership (affiliation network); (iii) based on the attribute of a specific social and professional position or (iv) by the participation in an event (e.g. a conference) (Marsden, 1990; Laumann, Marsden and Prensky, 1992). The problem is partly comparable to the problem of defining the population for a regular econometric analysis. However, boundary specification is even more pertinent in social network analysis (SNA) since interdependencies between actors are explicitly modeled and analyzed. Arbitrarily left out actors may lead to skewed or artifactual results. Third, a cognitive effect may bias the analysis of networks which are based on questionnaires or interviews. People tend to keep regular structures well in mind while they are less attentive for irregular events. Freeman, Romney and Freeman (1987) find that reports on the attendance of persons at a particular event show a tendency to include persons that generally attended the event but not the particular event asked for, while, on the other hand, persons who attended the particular event but attended other events only irregularly were less remembered. The conclusion drawn from this example is that it is hard to report on interaction which takes place in a delimited time frame. Instead, it is relatively easy to remember regular enduring activities or persisting social relations (Marsden, 1990). Fourth, when designing a study one has to decide whether to investigate actually existing networks or if one is more interested in networks as they are perceived by the actors (also called cognitive networks) (Marsden, 1990). While for the study of attitudes and opinions perceived networks are more important, for the study of innovation diffusion processes or cooperation, knowledge about more formalized networks is appropriate.

Interaction is a prerequisite for a variety of exchange processes – e.g. exchange of knowledge or other resources – but also for influencing mutual behavior. When applying the approach of routinized ties the character of the network is rather static. The move to a dynamic analysis creates a need for additional information about the starting point and the end point of interaction. This problem can be solved in different ways: First, if there are formal limitations of interaction as defined in a contract, these dates can be taken as start and end date, even though in reality (informal) interaction starts before the official beginning of a common research project and probably exceeds the official finalization date. Second, an assumption about the typical duration of interaction can be made. In the case of co-patenting networks a time span of five years is often plausible. Third, in other cases such as friendship networks a start and end date has to be fixed according to the definition of the friendship concept or any other concept of social interaction which is applied (Marsden, 1990).

For the study delineated in this dissertation, I apply archival information from a German database called "Förderkatalog" to reconstruct interorganizational networks in the German automotive industry. The problem of many other methods of data collection, such as surveys, questionnaires or interviews, is that organizational representatives are not necessarily aware of (all) the existing relations to other organizations. This is presumably a more severe problem in larger organizations that employ specialized staff and organize into separate business divisions. While at least the formal relations can be more easily analyzed in smaller organizations, informal ties are even there difficult to identify from outside.

## 5.3    Social Network Analysis of Innovation Networks

Most networks which are meaningful to be analyzed from the perspective of an innovation economist, such as patent networks or R&D networks, easily encompass several hundreds of nodes and can thus hardly be analyzed by simple eyeballing. Social network analysis (SNA) can help us looking deeper into the structure of such networks. Newman (2003, p. 171) expresses the advantage of special statistical techniques by asserting that it can answer the question: „How can I tell what this network looks like, when I can't actually look at it?". Social network analysis constitutes a tool which enables us to describe the interaction structure of firms in a network. Moreover, it focuses on the varying positions of actors within the social structure and measures their degree of embeddedness. The resulting measures allow us to draw conclusions, for instance with regard to the advantageousness of a certain position. Moreover, social network analysis is applied as a first step to disentangle the complexity of the network architecture (Wasserman and Faust, 1994). However, as long as the dynamic characteristics are not taken adequately into account, we cannot understand the full complexity of networks. SNA-indicators have been developed which allow for a description of networks, their structures and to some extent their dynamics. With the help of SNA-indicators internetwork comparisons become possible.

For firms that are part of a network it is useful to understand the functioning of the whole network and the specific roles of particular actors. Social network analysis offers a toolkit that helps to gain such insights. Central questions which SNA for innovation networks seeks to answer are: Which network structures are supportive to (collective) innovation success? Which position in a network is advantageous for individual actors? How robust is a network structure and which roles in networks can we differentiate from each other? Indeed, when we analyze the architecture of an innovation network one cannot oversee that some actors are more central while others are rather located at the periphery, and some actor have many ties while others have only few. More central actors have a relatively better position to absorb knowledge and exert influence. Moreover, central actors are able to bridge between different knowledge fields and may even be able to hamper other firms from access to knowledge. In a nutshell, the analysis of the network architecture is telling us how the network is functioning. The following graphs (Figure 5) exemplify the link between structure und functioning:

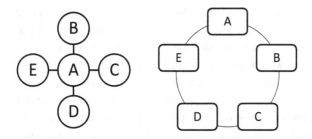

**Figure 5:** Star (left) and ring (right) network topology (Source: own illustration).

In the star structure, actor $A$ takes the most important position. This node has a link to all the other actors in the network, while the other actors can only indirectly initiate a link via actor $A$. Quite the opposite distribution of power is manifested in the ring network as there is no single central actor but all actors are linked to their immediate neighbors. The *degree centrality* which is measured as the number of direct ties between ego and the alters (Freeman Degree) has important implications for actors as well as for the entire network. For the case of innovation networks it can be understood as an indicator for the extent to which an actor gets the chance to access the network's knowledge base: The higher the measure is, the easier is the access. A couple of more complex indicators have been developed in order to describe the character of network embeddedness. For instance, the *closeness centrality* concept draws on the measurement of the length of path distances between actors and allows for a more comprehensive analysis of network structures (Jansen, 2006; Buchmann and Pyka, 2012a).

**5.3.1  Why Centrality Matters**

Nodes in general and firms in particular differ in their extent of network embeddedness. *Positional embeddedness* refers to the advantages with regard to knowledge access that is inferred from a specific position in the network. A central position in the network is supportive for innovative success (Ibarra, 1993; Ahuja, 2000a). It makes actors in a network both, more visible and attractive to be selected as partners and at the same time more influential without implementing more formal ways of control such as holding shares or having a formal say in the management board (Robinson and Stuart, 2002). The *degree centrality* is a measure for the direct connectedness of a firm to all other firms in the network. A high value indicates that a company is highly connected. Actors with high degrees of centrality exert power in the sense that they can control and brokerage knowledge in the network (Knoke and Yang, 2008). From a strategic point of view, not only the number of links to other firms is relevant but also the distance to other firms. The *closeness centrality* indicator captures the distance of an actor to all the other actors in a network. It gives an idea of how easy and quickly an actor can get in touch with other actors in the network directly or via only few steps (actors) in between (Knoke and Yang, 2008). If an actor has a central position which allows it to control knowledge flows between other actors in the network, then its *betweenness centrality* is high. That is, the more often an actor is located on the shortest path between other actors, the higher is the potential to control or moderate flows of knowledge and other resources, and to play the role of a broker or gatekeeper (Knoke and Yang, 2008).[6]

In addition, more central actors have a more comprehensive picture of the state of the network and more qualitative information about potential partners which reduces uncertainty about partner selection (Gulati and Gargiulo, 1999). Gilsing et al. (2008) find that the value of centrality is also a function of the technological distance between ego and the alter actor as well as of network density. For instance, in the case of average technological distances, more central firms which are embedded in fairly dense networks perform better when it comes to the development of explorative innovations. Note that the quality of calculations of network measures, e.g. degree distributions, depends on the completeness of the available data set (Powell et al., 2005).

---

[6] For a comprehensive overview of the formal centrality definitions see for instance Wasserman and Faust (1994) or Knoke and Yang (2008).

An effective strategy for a firm to attain a central position is to develop strong research skills that are appreciated by their partners and make these skills available to the network. Firms that are not conducting high-level research, which would allow them to maintain or expand their network, lose easily a central position (Powell, Koput and Smith-Doerr, 1996). In particular, Powell, Koput and Smith-Doerr (1996) find with regard to centrality: First, R&D cooperation influences positively the degree centrality. Second, collaborative R&D experience has a positive effect on closeness centrality. Third, non-R&D network experience affects both, degree and closeness centrality, and fourth, portfolio diversity has a positive influence on all three measures of central connectivity. It is the firms which gain the most central positions that provide impetus for the industry development (Powell et al., 2005).

Degree centrality also seems to be a good predictor for the growth rate of a firm. Size is the result of a company's behavior in the network, that is, growth processes are triggered by cooperative projects and reinforced by centrality. While cooperation leads to centrality, it is also centrality that affects in turn the extent to which a company is able to benefit from the network. Centrality has indeed two effects: First, it provides firms with access to a core group and thus enables them to benefit from critical knowledge and other resources. Second, centrality triggers a feedback loop as it supports firms with the formation of new ties and with the reinforcement of old ties (Powell, Koput and Smith-Doerr, 1996).

Centrality serves as a tool which enables firms to benefit from information asymmetries (Baum, Shipilov and Rowley, 2003). Central firms can easily find partners as centrality is linked to reputation (Podolny, 1993). Despite the increased freedom of choice with regard to a partner, there are hints that central firms prefer to collaborate with other central firms for different reasons: There is the principle of status homophily which suggests that firms with similar centrality levels are more likely to cooperate due the signal they emit with their position (Lazarsfeld and Merton, 1954; Podolny, 1994). Moreover, a firm might damage its own reputation if it cooperates with a less central firm (Podolny, 1993; Gulati and Gargiulo, 1999).

### 5.3.2  Indirect Ties

Not only direct links but also the indirect links play a role, reflected by the so-called *structural embeddedness*. Global structural embeddedness is formally expressed by the density of a network (Burt, 1995). Density measures to which extent the firms in a network are connected among themselves. In other words, it describes the realized links as a proportion to all possible linkages. For example, a network density which reaches a level of 21% means that 21% of all possible ties are actually established. Comparisons between network densities are rather difficult, as the number of nodes is negatively correlated with network density (Scott, 2000). Ahuja (2000a) suggests an interpretation of indirect ties according to which they constitute a way to maximize the benefits that can be drawn from the network. However, high network density is not advantageous per se. In cases when a firm seeks to access new knowledge, many close linkages make redundancy of knowledge very likely. On the other side, Coleman (1988) suggests that a dense network enhances the creation of trust and shared social norms which are both conducive for voluntary knowledge transfer. Oliver (2001) shows that firms acting in densely connected networks develop similar expectations concerning the behavior of other network participants, i.e. a system of norms and mutual control emerges. Rowley, Behrens and Krackhardt (2000) find that this allows for sanctioning actors not sticking to the commonly introduced rules.

Despite the apparent value of indirect ties, there is no mandate to explicitly foster the formation of indirect ties. Especially the replacement of direct ties with indirect ties cannot be recommended for the following reasons: First, direct and indirect ties can have a very different function for a firm. While direct ties provide resource-sharing and knowledge- spillover benefits, indirect ties are less relevant for the resource aspect which makes them less interchangeable. Second, there may be cases when both kinds of ties provide the same kinds of benefits. Even in these cases, as suggested by Ahuja (2000a), the benefits provided from indirect ties are relatively low as compared to direct ties.

Joining an innovation network will feed back on the behavior of other network members. Social norms of behavior are likely to emerge in densely connected innovation networks. With high network density a specific culture evolves framing the way actors think and act which distinguishes insiders from outsiders. Additionally, the costs for membership in networks are linked to the degree of density. For insiders of dense networks, costs can become low as the level of trust is expected to be high. Since the formation of network linkages requires a considerable investment and switching costs from one network to another are high, there should be no interest for opportunistic behavior.

### 5.3.3 Strong and Weak Ties

Participants in innovation networks benefit from their relational and structural embeddedness. Whereas strong ties facilitate the exchange of complex and tacit knowledge due to the possibilities of further inquiries, weak ties enable the network actors to access entirely new knowledge. Weak ties connect actors to remote subgroups in a network where – with a higher probability – rather new knowledge is located (Granovetter, 1973; Rowley, Behrens and Krackhardt, 2000). From a slightly different perspective we can also say that a strong tie network which consists of a lot of redundant ties is conducive to the diffusion of existing knowledge. The transfer of tacit knowledge is also accelerated in strong tie networks since redundant ties are an indicator for trustworthiness in the network. On the other hand, weak tie networks are more beneficial for explorative tasks, i.e. the generation of new knowledge which is restricted in dense networks in which redundant knowledge elements supersede (Rowley, Behrens and Krackhardt, 2000). The answer to the question which network features are beneficial differs from industry to industry. Rost (2011) analyzes networks in the German automotive industry with regard to structural and relational characteristics. She thereby shows that a combination of strong ties and low network density is most conducive for innovation.

### 5.3.4 Variety Regarding Exploration and Exploitation

A high activity level within a network makes a network for outsiders more attractive. Once a network attracts new firms it raises the variety of knowledge and thus the possibilities for new combinations. In other words, "diversity entails a preference for exploration over exploitation" (March, 1991, p. 79). Inside a network, diverse knowledge can often be reached by the formation of a tie to actors in more remote technological fields. The respective actors are often located in less well connected subgroups of a network. In network terminology, this means to bridge so-called *structural holes* (Burt, 1995; Burt, 2004; Powell et al., 2005).

The *structural hole* concept opposes the *social capital* concept. Social capital is approximated by the frequency of social interactions in a network. It refers to aspects of collective action such as trust and a system of values that determine the possible intensity of cooperation (e.g. Coleman, 1988; Burt, 1995; Walker, Kogut and Shan, 1997). The amount of social capital is determined by the quantity of network resources invested by a network actor, which in turn influences the space of opportunities and therefore has a strategic importance (Vonortas, 2009). However, Walker, Kogut and Shan (1997) find that the most beneficial network positions are those which enable the bridging of structural holes favoring more loosely knit networks. If we combine these considerations with growth phases of firms, it can be stated that, in the beginning, a dense network structure is more important while at a later stage bridging structural holes is more promising for a firm (Hite and Hesterly, 2001).

The self-structuring process of a network is influenced by the characteristics of an industry and by knowledge dynamics. Stable network structures very likely reduce knowledge variety. Mature innovation networks are particularly affected by this rigidity. Consequently, the major advantage of innovation networks, namely cumulative knowledge creation based on the variety of knowledge of network participants, might be sacrificed for the sake of specialization and a lock-in to relatively predictable technological trajectories (Kogut, 2000; Vonortas, 2009). Related to this point, Walker, Kogut and Shan (1997) detect a trade-off between stability and variety in network structures. The question arises whether there is an optimal structure balancing the two diverging tendencies, i.e. stability and variety. On the one hand, ties in an innovation network are established to connect formerly not connected knowledge areas, which give access to so far unexploited knowledge from cross-fertilization, i.e. bridging structural holes. Innovations networks of this kind aim at the exploration of the knowledge space. On the other hand, network links are established to better exploit the techno-economic opportunities of a specific knowledge area. Efficient exploitation is based on experienced practices while exploration is a routine changing activity itself. For the transition from exploration to exploitation, a combination of explorative and exploitative elements is conducive. For instance, a core group of actors doing exploitative work may link to surrounding actors or networks to access new ideas (Vonortas, 2009). Moreover, networks often show a clear division between a core group and a periphery group, and a concentration of knowledge transfer within the core group. These considerations also refer to Granovetter's (1973) concept of the *strength of weak ties* suggesting that weak ties avoid redundancy in the network and enable access to novelties.

Whether a rather lose network with flexible structures or a more dense network with well-rehearsed routines is more advantageous depends on the actual problem that an innovation network is confronted with. The idea of an equilibrated network structure might be misleading as it does not respect the two diverging objectives exploration and exploitation. Rowley, Behrens and Krackhardt (2000) reckon that high density and strong ties are better conditions for exploitation while low density and weak ties support exploration. In the same vein, Nooteboom and Gilsing (2004) suggest that new knowledge can best be discovered in structures of loose ties, whereas the transfer of complex and tacit knowledge requires dense networks. Hagedoorn and Duysters (2002) do not share this opinion and claim that the propensity to search for more radical innovations and to learn by exchanging knowledge with network partners increases with network density. They argue that for bounded rational firms acting in a permanently changing environment, connections between remote areas of the network are not of high relevance. Instead, the promotion of openness, network density and tie redundancy is supposed to be more effective.

As such, we have seen that there is no "one structure fits all network", but network structures are ideally tailored to specific purposes – even though the ideal pattern is not always obvious. Take for instance the case that all network actors are faced with similar new technological opportunities. In this case redundant interlocking ties are beneficial since they allow for the establishment of a high level of trust (Almeida and Kogut, 1999). Conversely, for a network that is more dependent on external knowledge or the brokerage of technology, loose and non-overlapping ties are more advantageous (Arndt and Sternberg, 2000).

### 5.3.5 Roles and Knowledge Absorption

Empirical evidence indicates that there is a significant stable relationship between different forms of centrality and the absorptive capacity (Cohen and Levinthal, 1990). Firms with low absorptive capacities get isolated in a network because their cognitive distance to other firms becomes too large to understand what these firms are doing. In contrast, once a firm has reached a comparatively high level of absorptive capacity, it is more likely to have many links in the network. Giuliani and Bell (2005) study network structures in a Chilean wine cluster by conducting a social network analysis based on interviews. Among others, they find evidence for the hypothesis that firms with a higher absorptive capacity have a higher probability to start a new tie. Furthermore, on the basis of their findings, three main roles of firms in innovation networks are conceptualized:

• Technological gatekeepers: These actors have a high degree of centrality in the network, that is, they are well connected within the network and they are also strongly connected with external sources of knowledge. They are hence a main knowledge source for other actors in the network and control knowledge flows. Concerning the strategic position, the role of gatekeepers entails many opportunities for firms in an innovation network: Bringing-in new external knowledge refreshes the innovation activities in the network and may re-focus the activities from exploitation towards exploration.

- Isolated firms: These actors benefit only rudimentarily from the network as they have only few links. From a strategic point of view, this is in general not a desirable position as isolated firms cannot influence the direction and intensity of innovation activities.

- External stars: Such actors possess strong linkages with external sources but only weak ties inside the network. These intra-network ties are almost exclusively focused on knowledge absorption. The role of external stars becomes visible in innovation networks that are composed out of firms characterized by very different sizes. In biopharmaceutical industries, for instance, the large pharmaceutical companies are typically connected with various small firms that conduct research in different areas of the knowledge space.

Only the technological gatekeepers have strong connections inside the network and can, thus, stimulate the learning dynamics with "fresh" external knowledge. They constitute the central element of a network's absorptive capacity. Strong external links enable these firms to enlarge their own knowledge-base and to improve their competitive strength. Less connected actors become aware of this disparity and try to connect to actors that are stronger than they are themselves, a process which increases the tie concentration (*preferential attachment*). Actors which want to play a gatekeeper's role and thereby strengthen the absorptive capacity of the entire network generally follow a process that consists of three steps: First, they gain access to external knowledge. Second, thereby they create new combinations of knowledge which is exchanged with other members of their network. Third, by exchanging newly acquired knowledge they foster the intra-network knowledge diffusion process. The voluntary transfer of knowledge without necessarily expecting reciprocal transfers is notable. The reason for this behavior is not altruism but the expectation of positive externalities, i.e. an improvement of the reputation of the entire network through improved products and processes.

## 5.4    Small-World Networks

Besides the actor related properties, another feature describing the architecture of networks becomes increasingly studied (e.g. Baum, Shipilov and Rowley, 2003; Hidalgo et al., 2007). A pervasive feature of many observed networks is (i) the formation of network subgroups which are very tightly connected and of (ii) loosely interconnected subgroups (Gulati and Gargiulo, 1999). This observation is in line with the concept of *small-world* networks (Watts and Strogatz, 1998; Newman, 2003). The small-world attribute is assigned to a network if it is significantly more clustered than a random network, and if the average path length is relatively small. The path length is defined as the average number of ties that lie along the shortest path between two nodes. It is a measure for the global structure of a network. In contrast, the degree of clustering measured by the clustering coefficient is characteristic for a local network subunit. For its calculation, the number of ties among all partners of a firm $i$ gets divided by the number of all possible ties that could exist among the partners of $i$ (Watts, 1999). To decide whether a network can be considered a small-world network, the values of the calculated (average) shortest path length $PL$ and the clustering coefficient $CC$ are compared with the values calculated for random networks with the same number of nodes and ties.[7] Yet, there is no generally accepted consensus with regard to critical values for these parameters. In large networks with high levels of clustering, it is often sufficient to establish some shortcuts between remote actors in order to fulfill the prerequisites of a small-world network (Watts and Strogatz, 1998). Innovation networks having a small-world architecture combine two advantages: (i) faster knowledge diffusion compared to random and regular structures and (ii) provision of new knowledge to all network participants (Cowan and Jonard, 2003).

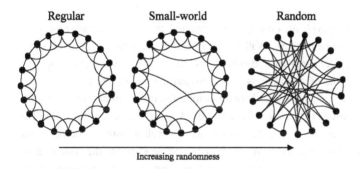

**Figure 6:** Network topologies (Source: Watts and Strogatz, 1998).

Three "archetypes" of network structures are identified, namely the regular, small-world and random structure (Watts and Strogatz, 1998) (Figure 6). These networks range from a well ordered, regular structure linking each node with its immediate neighbor (left network in Figure 6) to a random structures linking nodes arbitrarily (right network in Figure 6). In a regular network, the average path length $PL$ increases linearly with the number of nodes $n$. This development leads to a high degree of

---

[7] In a simplified way, the clustering coefficient can be calculated for a random network as $k/n$ and the path length as $ln(n)/ln(k)$ (Watts, 1999). $n$ denotes the number of nodes and $k$ the number of ties.

clustering. At the other side of the range (random), the degree of clustering becomes relatively low as $PL$ grows only logarithmically with $n$, but the average path length is relatively short. In between is the small-world network which has a short average path length due to some shortcuts and a relatively high degree of clustering.

When we observe small-world network structures, the questions about the emergence of this peculiar structure arises. Are small-world characteristics really a natural property of all networks, including interfirm networks? According to Watts (1999), small-worlds appear coincidentally in sociological, biological and technological networks. They are said to emerge from arbitrary interactions among actors in a network. In contrast, in the case of interfirm networks it could be the strategic behavior of firms which leads to a small-world structure rather than arbitrary interaction (Baum, Shipilov and Rowley, 2003). Baum, Shipilov and Rowley (2003, p. 698) assert that "there is a lack of empirical research exploring the evolution of small-world structures over time". In particular, the questions about the emergence and persistence are still underresearched. There is especially a lack of (empirical) explorative work and evidence about how behavior influences the emergence of a particular network structure.

In networks, firms are typically only loosely connected, that is, only a small proportion of all possible ties are realized (Walker, Kogut and Shan, 1997). Moreover, dense subgroups of firms (cliques) emerge (Gulati and Gargiulo, 1999). These properties explain a high degree of cliquishness. Ties between the cliques are responsible for considerably shortening path lengths between actors. The drivers of network evolution, such as triadic closure (cf. chapter 7), leads do dense cohesive substructures. The following question arises: Where do these cliques as well as inter-clique ties come from? The strive for embeddedness provides an explanation for the formation of cohesive subgroups, but the short average path length between the actors can only be realized if ties between clusters emerge. Thus, the question is whether some form of strategic behavior explains the establishment of such ties. One possibility for explaining inter-clique ties is information strategies. According to Moldoveanu, Baum and Rowley (2003), a cooperative information strategy can be distinguished from a competitive information strategy. The key difference is distribution and commonality of knowledge in the network. The first strategy aims at distributing and spreading private knowledge with the objective to improve coordination and common action within the clique or the network, subsequently establishing highly interconnected structures. This is advantageous because coordination is improved through repeated exchange with stable partnerships that facilitate the transfer of tacit knowledge

(Coleman, 1988). Whereas, with a rather competitive strategy a firm seeks to moderate and control knowledge diffusion in the network in order to gain a high betweenness centrality and to increase its influence and power. One possibility to implement such a strategy is to occupy structural holes in the network (Burt, 1995). By following this strategy, firms position themselves in so far unconnected parts of a network and bridge between independent subgroups. This gives a firm the chance to broker and control knowledge between two subgroups.

## 5.5    Conclusions

The collection of network data to be analyzed is particularly challenging in the case of longitudinal studies and large sample sizes. Social network analysis (SNA) is a valuable tool to analyze network structures and to identify characteristics of actors (position, role etc.) and of ties. Moreover, beneficial positions and ties in one situation may be inappropriate in another situation. For instance, exploration may require a larger diversity (and less density) of actors to have greater combinatorial possibilities of knowledge. On the other hand, strong ties might be important to build a high level of trust which facilitates the transfer of tacit knowledge. To analyze and assess the features of (innovation) networks, inspiration can be drawn from networks in other fields, such as biology or physics. There are obviously uniform laws governing network structures in many fields. More research is needed to identify potential causes for the emergence of such laws (see for instance Müller, Buchmann and Kudic (2013) for an investigation of the causes for the emergence of small-world networks).

# 6.    Towards a Dynamic Framework of Network Analysis

"Classic" cross-sectional network analysis, as presented in chapter 5, captures the development state of a network at a certain point in time, i.e. it takes a snapshot of the status quo. This allows for studying the structural characteristics of the network and the positioning and embeddedness of individual actors. There is meanwhile a large body of established literature which applies this type of analysis, but there are also shortcomings with this approach. Even though we may conclude from SNA in combination with a correlation analysis that a certain network related correlation exists, the direction of causality is often very ambiguous. Take for instance the case of a hypothesized beneficial position in a network and an observed higher rate of innovation generation. Is it always the position that drives innovation (behavioral dynamics), is it rather the success and potential to develop good ideas that brings firms in a central position (network dynamics) or do both effects play a role and even influence each other (co-evolution)? Such a question can hardly be answered with cross-sectional data but requires access to longitudinal data to give a more profound answer. An interesting and for economic research valuable part of the analysis is to detect not only change processes as such, but to identify and capture driving forces and patterns that are present in the observed networks. Borch and Arthur (1995) criticize that researchers collect too often data only for a limited time period, even though their ambition is to investigate the dynamic aspects of network evolution. Consequently, the conducted studies usually have the character of cross-sectional analyses and the developed models lack important dynamic aspects. The existence of complex interaction patterns is probably one of the reasons why the evolution of complex adaptive systems (CAS), such as networks, is not sufficiently studied within economics. Often, it simplifies the analysis by focusing on patterns in a state of „behavioral equilibrium", i.e. static patterns that are simplified aggregates of firm behavior (Arthur, 1999).

## 6.1    Networks as Complex Adaptive Systems

Complexity science regards socio-economic systems "not as deterministic, predictable, and mechanistic but as process dependent, organic, and always evolving" (Arthur, 1999, p. 109). Economic actors continuously adapt their behavior on markets, such as purchasing decisions, prices and expectations. Strategic action, foresight and adaptiveness make the development of adequate models challenging. In particular, the dynamic character of innovation processes requires us to take these features into account and go beyond mainstream economic analysis which studies states of behavioral equilibrium without considering further actions or reactions over time (Arthur, 1999). Schumpeter associated the idea of an evolutionary development with the concept of an economic sociology taking into account the endogenously changing framework of developing institutions. While changes in capital, labor or technology may lead to rather simple reactions of the economy, it is the "human factors" working on innovations that are responsible for the complex evolution of the economy (Schumpeter, 1911; Shionoya, 2007).

### 6.1.1 Complex Adaptive Systems

Due to dynamics, interaction patterns and adaptive behavior of firms, innovation networks are assigned to the comprehensive and general class of *complex adaptive systems* (CAS) (Holland, 1995). The constituent parts of a CAS, i.e. the actors, are not homogenous but can be distinguished in important characteristics from one another. Change in the system is driven by interaction of actors. It does neither follow a simple linear trend nor is it fully chaotic. Potential non-linearities in CAS signify that the behavior of a system "is more than a simple sum of the behaviors of its parts" (Holland, 1995, p. 5). Arguing in this vein implies that equilibrium models or simple trend analysis techniques can yield misleading results. Firms do neither interact randomly with all other actors in their environment in a way the molecules of gases do, nor do they interact solely with their immediate neighbors such as magnetic spins in a lattice. Instead, each firm interacts with a relatively small number of other firms. However, the interaction is in its resulting effect not limited to actors directly involved, but exerts influence throughout the network (Barabasi, 2007).

A further element of CAS is adaptive behavior of actors. Adaptation is, first of all, a typical reaction of organisms when they are faced with competitive pressure. However, the notion of adaptation – as it is generally understood – is too narrow to describe the actual process of selection. Adaptation refers in most cases only to "any feature that promotes fitness and was built by selection for its current role" (Gould and Vrba, 1982, p. 6). Besides, there is a second concept, namely *exaptation*, which plays likewise an important role. It describes characteristics which "evolved for other uses (or for no function at all) and that were later "coopeted" for their current role" (Gould and Vrba, 1982, p. 6). Such exaptations evolve from a combination of micro and macro constraints and processes. Considering both concepts, adaptation and exaptation, enriches the analysis of economic micro actors. In fact, there are applications of instruments, technologies and skills that were originally developed for a different purpose. For instance, the famous yellow post-it-notes were developed by the company 3M out of a mistake when they were looking for a new super-strong adhesive. What they discovered was however an adhesive which is weak from the beginning and does not get harder or softer over time (Brand, 1998).

There are more characteristics and mechanisms which are on a very general level common to all complex systems that are built from (adaptive) interacting parts (Holland, 1995):

- First and most straightforward, aggregation of system elements is based on categorization putting elements which share the same category in the same basket. This implies that innovation networks are based on the interaction of firms belonging to a network structure.
- Second, many concepts in economics assume linearity in causes and effects and consequently linearity in the models. Linearity means that the value of a function is a weighted sum of the values of the independent factors in the model. However, complex systems do not necessarily behave according to this strict assumption of linearity, instead they allow for nonlinear patterns that make predictions a lot more ambiguous (cf. chapter 6.2).
- Third, CAS change over time and have a flow character. They consist of three principle elements, namely nodes (firms), ties (channels of interaction) and resources (knowledge) that are transferred. While the network changes, new nodes and ties can come into existence but also disappear if firms are not able to adapt. "Neither the flow nor the network are fixed in time. They are patterns that reflect changing adaptations as time elapses and experience accumulates" (Holland, 1995, p. 23).
- Fourth, diversity is a further property of CAS. Each firm finds its position in a network as a consequence of its interactions. If one actor is removed from the system it undergoes a process of re-configuration eventually replacing most of the interaction pattern of the removed actor. This resembles convergence processes in biology. Also, a change in focus of one actor provides new opportunities for other actors to change their interaction patterns in turn.

Two important mechanisms are immanent to CAS and can be found in (models) of innovation networks (Holland, 1995):

- A first one is *tagging* which enables firms to purposely select an interaction partner. Without tagging, partner selection would happen in a purely arbitrary manner. Thus, it is a prerequisite for any kind of strategic action and for the formation of persistent subgroups to being able to distinguish actors from each other.

- Second, actors process knowledge from outside based on an *internal model* which allows them to make expectations regarding the future. This modeling element, which has a tacit character, determines a current action of a firm by envisaging a target state of the future. The assumption that actors seek to attain a desired state of a system is – as we will see in chapter 9.5.1 – operationalized in the applied model by an objective function which firms try to maximize by a myopic stochastic optimization rule. The internal model is influenced by the perceived environment which changes over time. "Perceived" implies that firms reduce the complexity of the reality into more abstract blocks that reflect the focus of their perceptional capabilities.

### 6.1.2 Modeling Tie Relatedness

A major challenge for modeling complex network structures is endogeneity which leads to related ties and violates the assumption of independent events of many statistical analysis techniques. In networks, we observe endogenously developing dependencies like clustering or triadic closure (formation of substructures which consist of three nodes). In view of these characteristics, regression analysis techniques which require statistical independence of tie formation cannot be applied. An appropriate alternative are stochastic network models following the approach of explicitly modeling dependency (Snijders, 1996). Robins et al. (2007) provide a list of arguments why we preferably apply stochastic methods for the analysis of network dynamics:

- A "breeze" of randomness, even just a small one, into a generally regular network formation process can make it very difficult to predict the outcome (Watts, 1999). The advantage of a stochastic model is related to its ability to take both features into account, the regular as well as the random one.
- A statistical significance test can tell us which (social) processes most probably drive the evolution of a network. The formation of certain substructures can be identified and shows that it is not just a random evolution.
- The formation of a certain kind of substructure can be caused by a social process but also be driven by the properties of the agents. If we take for example the tendency of clustering of actors in a network, it could either be a structural effect, such as structural balance, or rather an effect on the node level, such as homophily,

leading to the very same result. If we include both effects in a model we can decide which one of them is more important in the network we observe.

• A central question in social network research is: How do the dynamic processes on a local level lead to the formation of substructures, and in turn, how do these subunits combine to build the overall network structure. In most cases this investigation cannot be done without a proper model since the interaction of subunits is complex and the outcome is not evident. The simulation approach of a stochastic model can help to understand the micro-macro link.

While lots of effort was spent during the last years to develop more sophisticated and fine grained techniques of (cross-sectional) social network analysis, some researches focused on network change aspects and developed methods for modeling network evolution processes based on empirical data. Prominent examples are the "Exponential Random Graph Model" (ERGM) (Robins et al., 2007) and the "stochastic actor-based model for network dynamics" (Snijders, 1996; Snijders, 2001; Snijders, 2005). ERGMs use as their core element a "probability mass function" (PMF) which specifies the probability that a random graph is drawn from the same distribution as an observed graph. In doing that, we are able to explore the foundations of a certain network structure. By feeding the model with observed data we can estimate model parameters and test the model goodness of fit. However, different processes can lead to similar structures and ERGMs cannot account for this. For example, a tendency for clustering is frequently observed in social networks and different micro processes can lead to the same pattern: (i) persons who are socially highly active create clusters; (ii) homophily tendencies leading to assortative mixing may result in clustered networks; (iii) triadic triangles create clusters. It is highly desirable that we are able to fit these effects simultaneously and disentangle underlying mechanisms which becomes possible with the "stochastic actor-based model for network dynamics".

## 6.2    Agent-Based Modeling

Capturing real-world dynamics of innovation networks requires a toolkit which allows for the explicit consideration of the rich dynamics of firm interaction and the heterogeneity of actors. Agent-based (simulation) models (also referred to as actor-based models) focus on micro mechanisms in a longitudinal analysis and fulfill these requirements as well as the ones of modeling CAS delineated in chapter 6.1.1. Agent-based modeling (ABM) focuses on the behavior of individual actors and thus on the lowest (meaningful) level of aggregation. "ABMs deal with the study of socioeconomic systems that can be properly conceptualized by means of a set of 'micro-macro' relationships" (Pyka and Fagiolo, 2005, p. 468). This focus enables us to model in a more realistic manner aggregate phenomena such as network formation and evolution or the diffusion of innovation that happen within a context which is characterized by interaction and mutual influence of actors. By focusing on the modeling of network actors, we assume that network evolution is based on individual decisions (which can be influenced by other actors).

A first series of economic simulation models was designed to model complex system interaction but was not linked satisfactorily to empirical phenomena. A second series, developed since the end of the 1990s, was better able to replicate empirical findings and stylized facts. Technically, simulation is "a set of 'laws' that relate the dynamics of the process to the progress of the calculation" (Holland, 1995, p. 145). A general guideline for (simulation) models is to keep modeling mechanisms rather simple in order to avoid an obviously pre-deterministic behavior of the model, also called *unwrapping*. Unwrapping takes place if the final result of a simulation model is built into the program code in a way that a certain result or path is quasi inevitable to emerge. In this case, the final result becomes simply stepwise revealed during the simulation run and it is not emerging from the interplay of micro mechanisms. Consequently, it cannot provide new insights and limits the scientific value of such a model considerably (Holland, 1995). A further challenge of modeling can be described as follows: On the one hand, a model should be general enough to reproduce a broad variety of different phenomena. On the other hand, models should be able to describe and explain very specific phenomena in all their richness (narrative models). This trade-off can – if not fully solved – at least be mitigated by the development and application of agent-based models (ABM) (Pyka and Fagiolo, 2005). On the micro level, ABMs are based on heterogeneous actors which interact with each other. Specific interaction patterns emerge over time. ABM aims at the description and analysis of such complexities based on the characteristics and behavior of micro actors. For instance, Müller, Buchmann and Kudic (2013) study the network structures which emerge if a certain partner selection strategy is applied as well as the relationship between network diversity and innovation performance. We analyze the structural consequences of homophily, reputation and cohesion mechanisms in a scenario of information scarcity. In this context, agent-based modeling addresses the emergence of macro structures from simple micro strategies, such as cooperation routines. Moreover, we illustrate (i) that a transitive closure mechanism combined with a tendency for preferential attachment produces networks that show both, small world characteristics as well as a power-law degree distribution; (ii) diversity in the selection of cooperation partners is important for the innovative performance.

Furthermore, complex network phenomena such as path dependencies, dynamic returns, emergence of structures as well as mutual knowledge generation and learning can be analyzed. ABMs of innovation networks are based on actors characterized by (i) imperfect knowledge and the aim to improve the knowledge-base, and (ii) on actors that are confronted with uncertainty. Knowledge imperfections and uncertainties are tackled with the help of cooperation partners. Firms can improve their innovation performance by increasing the size and quality of their knowledge-base through learning from others. In comparison to approaches of micro aggregation (microfoundational macro models), ABM shows two decisive advantages: First, ABMs are able to showcase processes of emerging collective phenomena. Interaction of agents can be analyzed in depth, including the causes and effects of individual behavior of heterogeneous agents and the magnitude of their contribution to collective phenomena. Also, the timely process when a system starts to destabilize or decay can be observed in a microscopic manner. Second, ABM not only allows to conduct in-depth analysis of complex systems, but by understanding the systemic forces it becomes possible to setup computational laboratories which provide an idea of how the systems could evolve considering a varying institutional framework or incentive structures established by policy makers (Pyka and Fagiolo, 2005).

The general approach of ABM is about designing a model bottom-up in an inductive manner, starting with heterogeneous actors and the description of their behavior and endowments. A key challenge is to strike a balance between a high level of descriptive accuracy and observable micro processes. In other words, rendering a model more realistic by introducing sophisticated behavior rules and interaction patterns raises the level of complexity which goes at the expense of understanding causal relationships and implications of parameter variation. Researchers who apply ABM need to be aware of this trade-off and have to decide which approach they find more fruitful for the problem to be analyzed (Pyka and Fagiolo, 2005). Additionally, mechanisms which cannot be captured with empirical data can be explored. *Computational laboratories* enable us to investigate complex systems more accurately than it would be possible with econometric techniques alone. In particular, they can demonstrate that a (micro) mechanism is sufficient to generate an observed phenomenon and deliver hints for situations when a system starts to develop into an unexpected direction (Holland, 1995). Saviotti (2009) points out that the important explananda in economic systems are (i) the emergence of order in an evolutionary process and (ii) possible discontinuous transitions a system may undergo.

During an evolutionary process, a system may reach a point from which it can further develop into many different directions. This point is called *bifurcation point* (Poincaré, 1885). An industry can have a rather stable development path for a long time and change only gradually up to a point when change in technology (e.g. a shift from the internal combustion engine to the electric automotive engine) may change the industry structure more radically. Some firms may be prepared or even drive the change while others are not able to adapt. The result is an abrupt change rather than a gradual transition and sudden increase in the number of possible system states (development paths). For the case of innovation networks, this means that the structure of the network and the relevance of certain attachment mechanisms may change substantially over time. For instance, the advent of a new technology may require a rapid shift towards more explorative R&D and consequently towards partners which have more dissimilar knowledge-bases. Accordingly, longitudinal models which seek to investigate the drivers of network dynamics and/or the drivers of behavioral changes should control for time heterogeneity in the estimated parameters. From a period of relative "stasis" an industry may undergo at a certain point a rapid change process, also described as punctuated equilibrium (see also chapter 3.3.4). Relatively small events can, via feedback loops, become that powerful that they determine the further system development path (e.g. the industry structure) in the future up to the next bifurcation point. Thus, even if we know all the determinants which influence the development of a system, for a system that can be described by a non-linear equation, small changes in initial conditions may lead to very different and hardly predictable outcomes over time. These considerations are reflected, for instance, by the *logistic map* (Figure 7) which shows the non-linear characteristic of a system that is described by a relatively simple equation. As an example, a simple model which is used to describe population growth is written as:

$$x_{n+1} = r \cdot x_n (1 - x_n) \quad (12)$$

$x_n$ can take values between 0 and 1 and informs about the ratio of the existing population size to the maximum possible size in year $n$. $r$ designates a combined rate of starvation and reproduction (e.g. firm leaving and entering an industry) thereby taking values larger than 0, and $x_0$ represents the initial ratio of population to the maximum population size in year 0 (Verhulst, 1845; Verhulst, 1847; Ricker, 1954; Briggs and Peat, 1990).

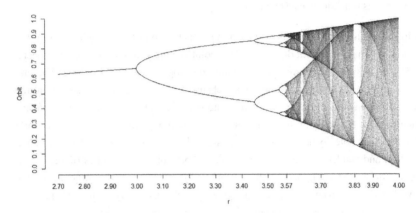

**Figure 7:** Logistic map (Source: own illustration, applied R-code adapted from: http://www.r-bloggers.com/logistic-map-feigenbaum-diagram, see C. Appendix).

The abscissa in Figure 7 shows the values of the parameter $r$. The ordinate shows the possible long-term values (after 1000 periods) of $x$ (accumulation points of the sequence; $x_0 = 0.1$). Small values of $r$ result in single stable fix point (convergence points). For $r = 3$ the first bifurcation point is reached and the graph is split up into two alternating streams of development. The two streams split into four streams once the next bifurcation point is reached. The tendency of a doubling at each bifurcation point and rather well defined lines continues (with decreasing parameter intervals) up to a value for $r$ of about 3.57 when the system starts to fall into chaos. Once the chaotic behavior has set on, a typical characteristic becomes revealed, namely very small changes in $r$ result in dramatically different results. Remarkably, in the sea of chaos we can find small islands of stability for instance at about $r = 3.83$ with an alternation between three values only. Thus, such a system is very sensitive for alternations in parameter values. This holds for most values between 3.57 and 4. Consequently, it becomes impossible to make predictions once $r$ grows beyond 3.57, i.e. predictions become in an exponential way progressively worse. For values of $r < 3.57$ there is an infinite number of fixed points with different periodicities and an infinite number of periodic cycles (Li and Yorke, 1975). Such a behavior entails problems with regard to models of socio-economic systems. Small errors in our knowledge of initial parameter values can lead to wrong predictions due to sequences of iterates that diverge exponentially. While exact predictions are impossible in non-linear systems, we can still calculate probabilities for a certain state if there is a phase of stability (*attractor*).

A probability measure will give the proportion of times in which the system is moving around within the region of the attractor. Non-linearity relates to dynamical trajectories which are indistinguishable from a stochastic process. This knowledge can be used to make decisions based on system states. In other words, the fact that we cannot predict an exact state does not mean that there is total randomness (May, 1976).

The extent to which a socio-economic system tends towards non-linear behavior or rather behaves in an exactly predictable way depends, among others, on the adaptive behavior of the actors. With their reactions they can weaken or amplify the tendency towards extreme outcomes. For instance, they can be more or less well capable of adapting to new (even endogenously created) technological paradigms depending on the absorptive capacity (Cohen and Levinthal, 1990), the network embeddedness, etc. Compared to the modeling of systems in natural science, adaptiveness can make outcomes in socio-economic systems more stable and predictable or even more chaotic and unpredictable.

## 6.3   Conclusions

In view of the fact that "a complete theory of complexity does not yet exist" (Barabasi, 2007, p. 33), agent-based models can improve our understanding of complex systems such as evolving networks. Moreover, ABM can be used to explore innovation policy and potential results of policy intervention. The general assumptions which are made for complex adaptive systems (CAS) can be transferred to networks and their constituent parts. Compared to cross-sectional analysis, longitudinal studies enable us to identify the drivers of network evolution and to study the mechanisms of change on the micro level based on the interactions of agents. When model parameters are estimated over longer periods of time, a check for time heterogeneity should be conducted to control for discontinuous shifts in development patterns of an industry which may change the rules of the game, such as attachment mechanisms in networks.

# 7.    Determinants of Network Evolution

Evolutionary thinking in a network context emerged only recently on the research agenda. For instance, Glückler (2007) addresses the question how tie selection constitutes an evolutionary process in networks. More precisely, he argues that network tie selection processes cause retention and variation within network structures. Hite (2008) presents an evolutionary multi-dimensional model of network change that explicitly considers micro level network change processes. Witt (2006) argues that selection processes are, in line with Neo-Schumpeterian approaches, constitutive for evolutionary economics. Studies on the evolution of network structures have typically focused on external factors (e.g. competitive pressure) as drivers for change processes. The focus on external determinants is, however, not sufficient to understand which partners are actually selected, according to which mechanisms and preferences. Gulati and Gargiulo (1999, p. 1440) conclude: "While exogenous factors may suffice to determine whether an organization should enter alliances, they may not provide enough cues to decide with whom to build those ties". For instance, firms with rather poor technological competencies are not considered to be attractive to connect with (Ahuja, 2000b). Partners in the automotive industry are often selected by their level of technical knowledge (Dilk et al., 2008).

A core research question deals thus with the determinants for the emergence and dissolution of a tie between actors. From a firm's perspective the question addresses the preference structures that determine the decision to cooperate and to select a cooperation partner. There are essentially three types of determinants (Figure 8): (i) firm characteristics (covariates), (ii) differences between firm characteristics (covariate (dis-)similarities) and (iii) preferences in network structure (endogenous factors). In other words, it is neither the characteristics of the firms alone, nor the relationship structure in isolation which are important for analyzing the evolution of networks, but it is a combination of the three mentioned elements which has to be taken into consideration (Garcia-Pont and Nohria, 2002). According to the concept of cooperation partner similarity (distance / proximity) (McPherson, Smith-Lovin and Cook, 2001), similar nodes have a higher probability to form a tie between each other compared to more dissimilar network actors. Similarity may refer to a variety of dimensions. Firms may be similar with regard to technological, organizational or financial characteristics, or even comparable in terms of reputation and status. For instance, Gulati (1995b) as well as Rothaermel and Boeker (2008) demonstrate that status similarity increases the rate of tie formations in interorganizational networks. Once a network has left the infant phase, it ripens and becomes structurally more and more differentiated; hence, it incorporates information about all other network actors (Gulati and Gargiulo, 1999). The network becomes not only a repository of knowledge but also of firm reliability and capability.

Firms collaborate for a variety of reasons among which learning and coping with uncertainty are important ones (cf. chapter 4). However, collaboration itself is a source of uncertainty with regard to the selection of the most appropriate partner. The selection process generates search costs and even if the "ideal" partner is found, the risk of opportunistic behavior cannot be fully eliminated (Gulati, 1995a). The central goal is to find a collaboration partner which complements a firm's own knowledge-base and at the same time is reliable and does not follow a hidden agenda (Van de Ven, 1976). This precondition is of special importance for the case of longer lasting collaboration endeavors such as collaborative R&D projects. Gulati and Gargiulo (1999) show that networks incorporate information about the complementarity or the reliability of a potential partner and are thus a source of relevant facts to make a partner choice. Firms may use this information, for instance, to minimize the risk of opportunistic behavior by considering the reputation of a potential partner. Reputation works as a social signal for firms in search of a cooperation partner and helps to select a potentially valuable partner (Dollinger, Golden and Saxton, 1997).

Derived from previous research in different fields, such as innovation economics, management science, economic geography and sociology, I introduce in the following subchapters effects which hypothetically play a strong role for the selection of cooperation partners in the empirically analyzed innovation network. I am particularly interested in knowledge related effects, following the knowledge-based approach described in chapter 4.2.3. These effects will, in a further step, be formulated as independent variables determining the evolution of the analyzed innovation network of German automotive firms. Figure 8 shows a conceptualization of the applied evolutionary network model.

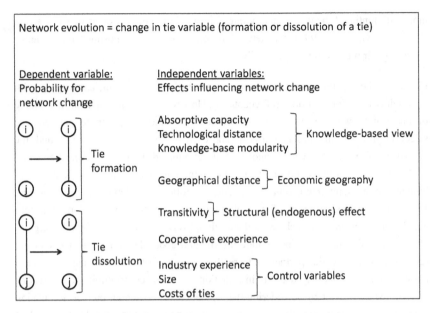

**Figure 8:** Conceptual model of network evolution (Source: own illustration).


## 7.1    Preferential Attachment

The simplest attachment model of a network is a random graph (Erdős and Rényi, 1960) which places (undirected) ties randomly between a certain number of $n$ nodes. Altogether, there are $n(n-1)/2$ possible ties in an undirected network with a probability $P$ (Newman, 2003). In contrast to the random attachment mechanism, Price (1976) finds that academic research papers which are relatively often cited, have higher probabilities to be citied again compared to less frequently cited papers. This "rich get richer" or *preferential attachment* mechanism is one of the essential non-random tie formation mechanisms. According to this principle, actors having many ties $k$ have a higher probability $P(k)$ to establish even more ties in the future compared to actors having fewer ties only. In other words, the probability that a new actor attaches to an incumbent actor in the network is proportional to his number of established links in the network, i.e. newcomers preferably connect to established actors that are already well connected. As a consequence, the distribution of the degree centrality in a network (the frequency of nodes with degree $k$) varies as a function of $\frac{1}{k^{\alpha}}$. This implies an inverse power law $P(k) \sim \frac{1}{k^{\alpha}}$, with alpha as the power coefficient. The graph of the distribution of degrees shows a long tail to the right meaning that there are many nodes with high degree levels (compared to a normal distribution). Whereas, in random networks (Erdős and Rényi, 1960) with large numbers of nodes, the degrees resemble a Poisson distribution (Barabasi, 2007).

Preferential attachment is regarded as an explanation for the development of *scale free* networks (Barabasi and Albert, 1999; Barabasi and Albert, 2002). However, Powell et al. (2005, p. 1137) criticize this "elegantly simple but over generalized" explanation for the phenomenon of scale-freeness. Other attachment processes are equally able to produce scale free degree distributions or even small world networks which represent network properties that can frequently be observed in real world networks (Müller, Buchmann and Kudic, 2013). Furthermore, Balland, De Vaan and Boschma (2012) find that the preferential attachment mechanism is strongly correlated with other structural attachment mechanisms and therefore decreases the goodness-of-fit level between simulated values and observed values in the stochastic actor based model for network evolution (Snijders, 1996). Consequently, I do not take the preferential attachment mechanism into account.

Economists and sociologists identified more mechanisms which are crucial for the development process of social and economic networks. In this regard, three general effects are relevant drivers of tie changes: First, the structural position of actors in a network plays a role, e.g. friends of friends become friends. Second, the characteristics of actors, e.g. the absorptive capacity (actor covariates), are determinants for the decision to collaborate. Third, characteristics of pairs of actors (dyadic covariates), e.g. their geographical distance, matter. The assumptions about potential drivers can be formulated as hypotheses. While classical social network analysis (SNA) is a type of descriptive network statistics, for hypothesis tests, an inferential-statistical approach has to be applied. This requires knowledge about distributions of test statistics for calculating *p-values* of hypotheses tests. The challenge is to construct distributions for entire networks (see also the permutation approach in chapter 9.3).

## 7.2    Absorptive Capacity

A first knowledge related independent variable to be considered is the absorptive capacity. In line with the knowledge-based view of the firm, firms differ in their ability to make use of external knowledge. This is substantiated by the concept of the absorptive capacity of a firm which reflects a firm's ability "to recognize the value of new, external knowledge, assimilate it, and apply it to commercial ends" (Cohen and Levinthal, 1990, p. 128). If a firm has already accumulated knowledge in the same or related fields, it is easier to recognize, evaluate, assimilate and apply external knowledge. Studies on the processes of learning suggest that storing new knowledge in memories is self-reinforcing, i.e. the more there is already stored the easier new knowledge can be acquired (Bower and Hilgard, 1981) and contextual knowledge is essential to make full use of new knowledge (Lindsay and Norman, 1977). In a nutshell: "Learning is cumulative, and learning performance is greatest when the object of learning is related to what is already known" (Cohen and Levinthal, 1990, p.131). Cumulativeness refers to knowledge creation in the present and future being a function of the knowledge with has been created in the past. Dean et al. (2012) argue that the success of human culture is predominantly related to the cumulative character of knowledge and technology over time. To identify the causes of cumulative culture they study social and cognitive capabilities of children, capuchin monkeys and chimpanzees. Their findings suggest that cumulative culture is unique to the human species. In fact, teaching, communication, observational learning and prosociality are found to be relevant elements of human culture only. Building up a highly effective absorptive capacity is not effortless but requires regular and intense examination and application of the existing knowledge-base to build associations between new and old knowledge elements (Lindsay and Norman, 1977).

Thus, firms have advantages in integrating and applying external knowledge when they can draw on own experience in research (Cohen and Levinthal, 1989). Moreover, we can distinguish between the prerequisites required by the receiver to make use of and integrate external knowledge, and the relevance of the characteristics of the external knowledge which will be discussed in the next chapter. The avenue of knowledge production can be directed by technological paradigms or by technological trajectories which reduce the degrees of freedom (Dosi and Nelson, 2010). Learning processes themselves may be improved by the development of learning skills which is referred to as *learning to learn* (Estes, 1970).

In networks, collaboration partners are faced with the risk of being "misunderstood" when they exchange knowledge due to a lack of absorptive capacity. In contrast to this, neoclassic economists stress a different risk, namely the risk of knowledge spillovers (at zero costs) (Griliches, 1992). Thus, a core proposition of the concept of absorptive capacity is that actors need strong internal capabilities in order to learn from external sources. With regard to the innovative performance of a firm, Tsai (2001) finds that the absorptive capacity together with the network position has a significant effect, and

Giuliani and Bell (2005) detect great variety and thus heterogeneity in firm absorptive capacities.

Cohen and Levinthal (1990) raise the question whether the absorptive capacity can be introduced to an organization from external sources via acquisitions, consultancy services or hired employees, or if it must be truly developed internally. They suggest that at least some elements of the absorptive capacity are very firm-specific which limits the possibilities of acquiring absorptive capacity from external sources. The successful integration of non-trivial knowledge presupposes, besides time, that a firm's engineers, technicians etc. are not only experts in their field but are also aware of the firm's internal organization, its routines, its external relations and operational procedures (Cohen and Levinthal, 1990). The very nature of knowledge may also require a firm to conduct basic research as it gives orientation to choose where and how to conduct more applied research downstream to the market once the most essential functioning of a technology or a natural phenomenon is understood. Basic research serves as a compass for applied research and to assess its possible consequences (Rosenberg, 1990). Progress in one technology may open new possibilities in another technology by cross-fertilization (Mokyr, 1990) which means that a firm needs a broad knowledge-base to master different technology fields (Cantner and Pyka, 1998).

### 7.2.1 The Value of Own Skills

Systematic incorporation of external knowledge is a prerequisite to survive in innovation competition. It allows firms to operate at the cutting edge of technology. However, the feature of permeable firm boundaries and innovation network embeddedness for internalizing external knowledge does not mean that internally created knowledge and own capabilities become obsolete. The opposite is actually true: Internal skills are a conditio sine qua non to detect, evaluate and integrate external knowledge. Effective exchange of knowledge requires a high level of absorptive capacity. It enables firms to understand what is going on outside their own organization. Firms which conduct own R&D are consequently more likely to absorb valuable external knowledge (Cohen and Levinthal, 1989). The incentive to invest in R&D and thereby to increase the absorptive capacity is dependent on the quantity of the knowledge that can potentially be absorbed and on the difficulty to absorb it. For the kind of knowledge which is more difficult to internalize, more prior knowledge needs to be accumulated which is more costly and requires higher prior investments in R&D. Firms that are operating in environments in which learning is relatively difficult, the costs for additional R&D are accordingly higher but have to be borne, otherwise firms will fall back in competition. Whereas, firms operating in fields where learning is relatively straightforward, the effect of R&D on the absorptive capacity is smaller (Cohen and Levinthal, 1990).

A network of collaboration is the instrument to access and to test R&D results in the community. For instance, Rosenberg (1990) studies basic research in the USA and finds that the majority of it is conducted within the university community. Firms which want to access and exploit this knowledge need highly developed capabilities and therefore have to conduct at least some own basic research in the respective field in order to absorb knowledge from universities. Giuliani and Bell (2005) demonstrate that knowledge in networks is not evenly distributed but often concentrated within a group of core firms that have significant over-average absorptive capacities. The extent to which a firm is able to learn new knowledge and apply it in a meaningful way is a function of the scale and scope of its previously acquired knowledge stock. If a firm has already accumulated knowledge in the same or related field, it is relatively unproblematic to understand related new knowledge (Frenken, Van Oort and Verburg, 2007; Boschma and Iammarino, 2009).

### 7.2.2 From Individual to Organizational Absorptive Capacity

The herein discussed firm models abstract from individuals and consider the organizational level. "An organization's absorptive capacity will depend on the absorptive capacities of its individual members" (Cohen and Levinthal, 1990, p. 131). It would be misleading to simply sum up the absorptive capacities of an organization's employees in order to determine the organization's absorptive capacity. Instead, the absorptive capacity of an organization is characterized by a variety of building blocks. Besides getting access to new sources of knowledge, an organization also needs to be able to meaningful use this knowledge. Therefore it needs the capability to internally process the knowledge which is at the moment of transfer extracted from the original context (Cohen and Levinthal, 1990).

Knowledge can effectively diffuse and be processed if there is some common understanding for it within an organization. On the other side, diversity between individuals instead of too much common understanding is beneficial as it forges the emergence of innovative solutions (Molina-Morales and Martínez-Fernández, 2009). Diversity enables the integration of different knowledge fields and the formation of new associations (Cantner and Pyka, 1998). Consequently, the absorptive capacity of a firm is not anchored in individual employees but "depends on the links across a mosaic of individual capabilities" (Cohen and Levinthal, 1990, p. 133). A firm which is highly specialized in one particular technology field has a very sophisticated knowledge in this field and is probably able to learn easily from related fields. However, the overall absorptive capacity of such a firm may still be rather limited since, due to the specialization, the number of related fields is rather small. Thus, it is advisable to develop a dispersed knowledge profile in order to cope with the uncertainty of selecting the right source of valuable knowledge. Diversity also works as a learning accelerator as it allows for bridging between knowledge fields (Cohen and Levinthal, 1990).

In line with Giuliani (2005), the absorptive capacity of a network of firms can be defined as the capacity of a network to absorb, diffuse and creatively exploit extra-network knowledge. Without a continuous integration of external knowledge, innovation networks might mutate to less innovative cliques as the knowledge of network members becomes increasingly homogenous after a sequence of mutual knowledge exchanges (cf. chapter 2.1). For this reason, the ability to refresh the innovation network's internal knowledge highly matters. The network absorptive capacity is a function of the member's capacity but not a simple aggregation as the links functioning as conduits of knowledge acquisition and diffusion are central components of the network capacity.

Previous research indicates that firms with higher absorptive capacities are better connected to sources of knowledge located outside the network. This can be explained with the size of a firm's individual knowledge-base that determines the possibilities to create links with external actors. Also, firms with a high level of absorptive capacity are cognitively closer to external knowledge and play the role of a knowledge gatekeeper which supports (or hinders) the diffusion of external knowledge inside the network. In other words, high absorptive capacities indicate that a firm's knowledge-base allows for more interfaces with the knowledge-bases outside the network which spurs knowledge transfer into the network (Cohen and Levinthal, 1990). By studying roles of network actors, Giuliani and Bell (2005) find that firms which have relatively higher absorptive capacities are more eager to build ties to sources of knowledge external to the network which is related to the observation that these firms are cognitively closer to external firms. Closeness facilitates the absorption of external knowledge. As a consequence, absorptive capacity can be regarded as a moderator which determines how much external knowledge can be transferred via the gatekeepers to the internal knowledge system.

**7.2.3   Interaction between Network Position and Absorptive Capacity**

The ability to internalize and apply external knowledge also influences the effect of the network position on the innovation performance. While a central position provides control and access to many sources of knowledge, the potential advantage can only be realized if the firm actually absorbs accessible knowledge. Firms with relatively low levels of absorptive capacity may still be able to spot interesting knowledge but they are unable to transfer it and internalize it. This is what Hansen (1999) calls a "search-transfer problem". Moreover, central firms have the advantage of being able to access a much broader variety of sources of knowledge compared to less connected firms. This advantage can however only be exploited if the central firm has developed the absorptive capacity that enables it to make use of the entire variety of knowledge. Consequently, a firm needs to invest in parallel into its absorptive capacity when it increases the number of ties in a network to make effective use of new ties. This process is not only time consuming but also costly which – in tandem with network administrative costs – sometimes shrinks the potential benefits of a large number of ties considerably (Tsai, 2001). Accordingly, I control in the applied network evolution model for the degree (density) which reflects the costs for additional ties.

**7.2.4   Path Dependency and Absorptive Capacity**

The absorptive capacity of a firm is to a large extent determined by the amount of previously acquired knowledge. The role of previous knowledge is indeed twofold and has to be analyzed over time: First, the level of a firm's absorptive capacity in $t$ is a determinant for the level it can reach in $t+1$. Second, as technology-based firms operate in highly uncertain environments, a broad knowledge-base helps to evaluate the significance of small changes in technology development. This is helpful to recognize at an early stage potential trajectories a technology may follow and to assess its potential for commercialization. The two effects imply in tandem that the development of the absorptive capacity is path dependent and influenced by historical states. A lack of investment into the right technology at an early stage can become a serious problem since there is a risk of missing the train and of never catching up again. The firm which is lacking behind at an early stage gets locked out of recent developments (Cohen and Levinthal, 1990).

Cohen and Levinthal (1990) suggest that the aspiration level of a firm in a certain technology field is dependent on the absorptive capacity rather than on the past performance. The more absorptive capacity it has developed, the more opportunities will be revealed to the firm and the more actively it will search for new technological and business opportunities. The interconnection between a firm's aspiration level and its absorptive capacity can result in a self reinforcing-cycle which keeps some firms persistently in a leading position while others are deemed to remain technologically-wise left behind.

## 7.3 Technological Distance

Lane and Lubatkin (1998) identify three distinct methods of knowledge acquisition, namely passive, active and interactive. The first form of learning, which is passive learning, takes place when people learn from written or oral contexts such as books, journals, seminars or consultants. Active learning, the second option, encompasses for example benchmarking and competitor intelligence which provides insights into a third firm's capability portfolio. However, only what can be observed or what is organizational intelligence can be acquired. Both described forms of learning can add only relatively unspecific and broadly diffused knowledge to a firm's knowledge-base. Such knowledge is known to a large audience and is therefore not scarce or costly to imitate and thus cannot be regarded as unique knowledge providing an advantage in competition (Spender, 1996). Interactive learning, which constitutes the third way of learning, enables a firm to acquire knowledge which has the potential to really make a difference. For this, the learning firm may be required to be located in short geographic and especially in short technological distance to a teaching firm in order to grasp the tacit elements of production or managerial processes. The transfer of this "how and why knowledge" requires a high level of trust between actors. It is context specific and hard to imitate, thereby adding real value which can in the best case eventually be transformed into profits (Spender, 1996). The required level of proximity (in various dimensions) can be realized by the formation of alliances of interacting organizations but not from simple observations.

Consequently, the capabilities to absorb and make use of external knowledge not only depend on prior R&D investments but also on the following points (Lane and Lubatkin, 1998):

- The similarity of the knowledge to be learned
- The similarity of the knowledge processing systems
- The similarity between firms' organizational structures and practices

That is, the ability of two firms to learn from each other is determined by dyadic firm characteristics. These aspects are reflected in the technological distance between two actors which is the second knowledge-related effect to be tested. As I focus on innovation networks, the similarity of the technological knowledge-bases is of outmost importance. The concept of technological distance refers to shared technological experiences and knowledge-bases (Knoben and Oerlemans, 2006). This understanding is somewhat similar to the concept of cognitive proximity (distance) as it is described for instance in Boschma (2005), even though cognitive proximity is more comprehensive.

The more the internal knowledge-base is related with external knowledge, the easier external knowledge can be captured. Consequently, a firm can learn more easily from other firms which belong to the same industry (Henderson and Cockburn, 1996) or which operate with similar technologies (Jaffe, 1986). For instance, in a descriptive manner, Yang, Phelps and Steensma (2010) refer to relatedness by conceptualizing a *spillover knowledge-pool* which consists of knowledge elements of the originating firm recombined with elements of the recipient firm and derivatives from them. To make meaningful use of the knowledge of other firms, external knowledge needs to be combined with the internal knowledge. Since knowledge-bases are heterogeneous, also the way in which firms can exploit external knowledge differs from firm to firm and is thus a distinguishing firm characteristic (Sorenson, Rivkin and Fleming, 2006). Thereof we can draw two conclusions: First, it is relatively easy to learn new things in fields in which we developed already some expertise, while it is relatively difficult in fields which are completely new to us. Second, the characteristic of a knowledge-base changes mostly incrementally due to the fact that learning takes preferably place in fields that are related and somewhat similar to familiar fields. It is not only the amount of previously acquired knowledge which determines the absorptive capacity but also the diversity (Cohen and Levinthal, 1990; Cantner and Pyka, 1998).

Lane and Lubatkin (1998) confirm for a sample of pharmaceutical-biotechnology R&D alliances that the similarity of the partners' knowledge-bases is positively correlated with interorganizational learning. In cases where knowledge is predominantly tacit, knowledge-base similarity in combination with strong ties is a necessary prerequisite for knowledge transfer. Knowledge that is transferred from one actor to another is always subject to the interpretation of the receiver, i.e. what is sent is never fully identical to what is received, which gives rise to misperception and misunderstanding. Consequently, with growing degrees of tacitness knowledge transfer becomes increasingly difficult. Learning necessitates a certain degree of similar problem perception (Cohen and Levinthal, 1990; Colombo, 2003). This idea transferred to the dyadic level suggests that cooperating firms must, for effective learning, have similar knowledge-bases which reflect a common understanding of technological problems. When the knowledge-bases are very dissimilar, firms are probably working on different technological problems and are following a different technological trajectory which means there is not much they can learn from their partner (Giuliani, 2010). Similarity in knowledge-bases facilitates communication, comprehensibility and thus the efficient exchange of knowledge. It supports learning and fosters the enlargement of a firm's own knowledge-base.

Note, however, that firms need to strike a balance between technological proximity and distance in order to guarantee a sufficient degree of novelty of the exchanged knowledge. Larger distance increases the probability for gaining access to substantially new knowledge with a potentially higher impact on innovation (Cohen and Levinthal, 1990) as invention and innovation are understood as new combinations of knowledge which requires more dissimilar knowledge-bases.

## 7.4 Knowledge-Base Modularity

As a third and – to the best of my knowledge – so far untested knowledge-related factor, I suggest the modularity of the knowledge-base to be a determinant of the preference to cooperate and select a partner. Modularity constitutes a basic evolutionary principle (Pyka, 2002). Complex systems often exhibit a modular structure, i.e. (i) interdependency between modules is low, (ii) interdependency within modules is high and (iii) modules can be reconfigured upholding the functionality of the system (Baldwin and Clark, 1997). Modularity has been mostly studied with regard to product architecture and organizational structures (see for instance Sanchez and Mahoney (1996); Baldwin and Clark (2000); Schilling (2000); Ethiraj and Levinthal (2004)). The purpose of a modular product architecture is to buffer elements of a system from each other and to prevent the emergence of ripple effects. A modular architecture strengthens the stability of the product and its performance as well as facilitates repair since modules can be replaced separately from one another. On the other hand, the ripple effects, which are harmful with products, are rather desirable in the context of knowledge since they trigger exploratory search between modules (Yayavaram and Ahuja, 2008).

Modularity effects, concerning the knowledge structure, have so far been of minor interest in innovation economics. A few studies identify a relation between the structure of the knowledge-base and innovation related outcomes. Lane and Lubatkin (1998), for instance, find that the degree to which two knowledge-bases overlap influences positively the ability of mutual learning in cooperation. Coupling of knowledge elements leading to a modular structure is related to three distinct motives: (i) there might be a natural interdependence between some knowledge elements; (ii) search routines may be directed to the coupling of certain knowledge elements while other elements are more independently used; (iii) innovation processes are recombinant which implies the coupling of so far unrelated knowledge elements. The varying degrees of decomposability explain why knowledge-bases that consist of the same knowledge elements may (and most often do) differ in their actual application (Yayavaram and Ahuja, 2008).

Brusoni and Prencipe (2001) summarize the suggestions made in the literature on modularity: First, there is a strong link between knowledge, product and organizational modularity which means that the knowledge encapsulated in a modular product is also modular. Moreover, there seems to be a link between the organizational structure of firm departments and the structure of the knowledge-base. A firm which conducts its R&D processes in sharply separated and independent working units is very likely to generate a relatively strong modularized knowledge-base (Yayavaram and Ahuja, 2008). Second, modular product architecture facilitates the division of labor inside a firm as well as between firms. Third, modular product architecture reduces coordination efforts in the division of labor. In particular, Arora, Gambardella and Rullani (1997) argue that modularity of knowledge and technologies supports the division of labor in innovative activities. In a stylized way, modularity divides innovation processes in two separate components, namely in (i) the production of new (basic) modules and (ii) their combination for tailor-made technologies and designs to meet market needs. While "specialized upstream suppliers" focus on the production of new modules stressing economies of scale, more downstream firms combine these modules to assemble complex products.

High levels of modularity often characterize mature industries such as car-manufacturing. Here, original equipment manufacturers (OEMs) outsource activities to increase the efficiency of manufacturing. They receive from their suppliers pre-assembled and pre-tested modules, e.g. doors, cockpits, etc. Whereas, firms operating in an industry that is still in an early phase of growth without a dominant design (Utterback, 1995), are more inclined towards vertical integration. From this follows that a modular production system is likely to stipulate knowledge division among firms. Suppliers develop and produce modules exploiting intensively their expert knowledge. For OEMs it is important to keep internally some knowledge of the modular content in order to integrate the outsourced modules into a final product (Takeishi, 2002). The global integration of production went hand in hand with an increased modularity of process technologies. This process gave external suppliers a more prominent role in the value chain taking over parts of the design and engineering tasks (Sturgeon, Van Biesebroeck and Gereffi, 2008).

Invention is often regarded as the outcome of knowledge recombination (Schumpeter, 1939; Ahuja and Katila, 2001; Fleming and Sorenson, 2001). Thus, firms not only try to find a partner which has a similar technological understanding, but they attempt to recombine and link technologies and the underlying knowledge. I expect firms which have modular knowledge-bases to be preferably chosen as collaboration partners because this facilitates the combination of knowledge. In particular, a decomposable knowledge-base enables researchers to conduct recombinant search processes without getting trapped in complexity and endless combinatorial possibilities (Yayavaram and Ahuja, 2008). The recombinatorial possibilities become rapidly very large even with rather modest sized knowledge-bases. Firms that search for an appropriate cooperation partner to combine elements of their own knowledge-base with elements or a partner's knowledge-base to come up with innovative solutions are confronted with a high level of complexity, an overload of possibilities and uncertainty at the same time. Hence, I propose that the propensity of two firms to cooperate rises with their ability to structure their knowledge-base in a modular way. Modularity reduces time and costly search process as compatible technologies can be identified more easily, and it decreases complexity through a reduced number of combinatorial possibilities. This argument relates to a study of Yang, Phelps and Steensma (2010) on 87 telecommunications equipment manufacturers. They find that the rate of innovation is higher if the external knowledge-pool (represented by patents) is greater and more related to the originator's own knowledge-base. This effect is yet not without limits. Once the size of the pool grows larger and larger, the positive effect shrinks and finally becomes negative due to overwhelming complexity which grows in parallel with a growing pool of knowledge.

The characteristic of a modular knowledge-base is synonymously also referred to as a clustered or highly decomposable knowledge-base. The feature of modularity of a firm's knowledge-base is approximated by its degree of clustering. The degree of decomposability is reflected by a continuum of structures. From a non-modular to highly modular a knowledge-base the ties between knowledge elements become increasingly clustered (Figure 9).

To summarize, firms with modular knowledge-bases are preferably chosen as collaboration partners because (i) it is easier to integrate their knowledge and (ii) it increases the speed of engineering of collaborative products. Accordingly, I suggest that not only the absorptive capacity or the technological distance matter for the propensity to cooperate but also that the decomposability of the knowledge-bases into modular knowledge substructures plays a significant role.

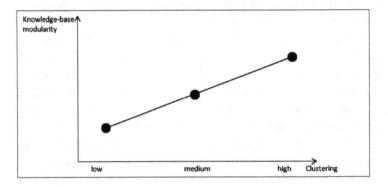

**Figure 9:** Stylized relation between modularity and clustering (Source: own illustration).

A highly modular knowledge-base is characterized by some knowledge elements forming a dense cluster while clusters are not knit together (Figure 10). Whereas, in nearly decomposable structures (Simon, 1962) nodes are clustered through dense links and there are links connecting the clusters (Figure 11). Finally, a non decomposable pattern does not show identifiable clusters but the ties are arbitrarily distributed (Figure 12).

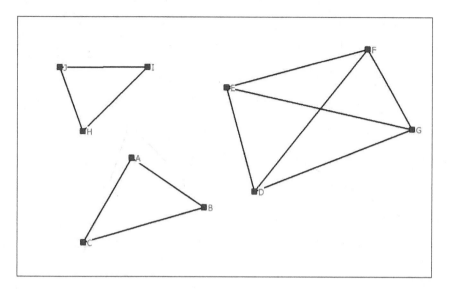

**Figure 10:** Modular knowledge-base (Source: own illustration).

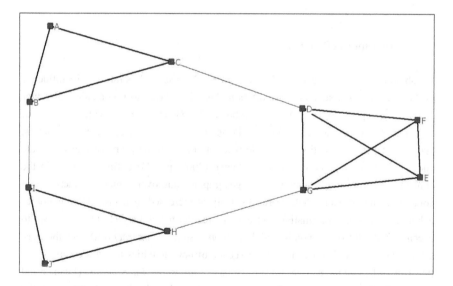

**Figure 11:** Nearly modular knowledge-base (Source: own illustration).

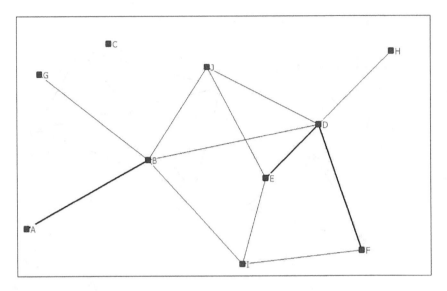

**Figure 12:** Non modular knowledge-base (Source: own illustration).

## 7.5    Geographical Distance

An obvious reason why geographical distance influences collaboration is its influence on transaction costs, such as transportation. Also, it is easier to find a suitable partner among a group of co-located firms (Tabuchi, 1998). Despite the wide diffusion of communication technologies which shrink perceived distances between actors, geographical distances still play a role when it comes to the propensity to cooperate and to select a cooperation partner. Feldman (2000, p. 373) defines location in the context of knowledge creation "as a geographic unit over which interaction and communication is facilitated, search intensity is increased, and task coordination is enhanced". Modern information and communication technology (ICT) has helped to decentralize a lot of economic activities (*conversational interactions*), but the very same technologies also lead to the emergence of new activities that are complex and thus cannot be conducted over longer distances (*handshake interactions*) (Leamer and Storper, 2001). Boschma and Wenting (2007) examine the local clustering of the British automobile industry in the period 1895-1968 and find two evolutionary explanations for concentration patterns, namely agglomeration economies in combination with spinoff dynamics. In regions, where a lot of firms from related industries were located, such as bicycle or coach making, the hazard rate of automotive firms was considerable lower. This holds particularly for the infant phase of the industry. Knowledge externalities and locally bound skilled labor are to a great deal the explanations for this observation. Even in highly computerized industries personal interaction is still inevitable. Tasks which cannot be standardized and routinized require a high level of implicit knowledge (Storper and Venables, 2004). In fact, in various industries we find tendencies for an uneven distribution of firms in geographical space which is a first indication that geography matters for firm location. This holds in particular for high-tech industries (Audretsch and Feldman, 1996). Accordingly, geographical distance is considered as a further determinant of network formation.

Innovation is a collective learning process which takes to a large extent place within a spatially bounded local milieu (Camagni, 1991) and many innovation networks have indeed a strong regional focus (Boschma, 2005). Geographic proximity facilitates the exchange of knowledge among firms. Jaffe, Trajtenberg and Henderson (1993) show that geographical proximity is conducive to mutual learning. Knowledge flows are typically stronger between close and similar regions or countries and weaker between regions that are further away from each other, that have different languages or are specialized in dissimilar sectors (Peri, 2005). The extent to which firms tend to network with other firms in close geographic distance depends among others on the degree of tacitness of the respective knowledge. The transfer of codified knowledge via journals, books, etc. is relatively independent from the geographical distance. In contrast, exchange of tacit knowledge requires personal contacts and trust which is easier to develop in a regional context (Von Hippel, 1994). Leamer and Storper (2001) demonstrate that the transfer of tacit knowledge via modern information and communication technologies is difficult. Jaffe (1989b) provides evidence for the effectiveness of spillovers by considering a regional parameter in the estimations.

However, I doubt that short distances per se improve the diffusion of knowledge. Rather, short distances facilitate the establishment of network ties which allow for extensive knowledge exchange and the collection of information about activities of co-located firms. The gist from different geographical observations and studies is that there are essentially two channels by which distances exert influence (cf. Glückler, 2007): First, short distances positively affect the formation of interfirm networks. It is not the physical distance as such which influences network formation. Instead, it is the possibilities and preferences of human beings to communicate (Storper and Venables, 2004), i.e. infrastructure and possibilities to travel faster are to be taken into account (Marquis, 2003). We often find not only a tendency for clustering with regard to an industry's location but also in terms of interaction patterns (Weterings, 2006; Hoekman, Frenken and Van Oort, 2009). Shorter distances provide more opportunities to meet which is conducive for developing trust serving as a precondition for knowledge exchange (Howells, 2002). Face-to-face interaction facilitates interactive learning. Thus, there is an indirect relationship between geographical distances and the possibilities and propensities to form fruitful agreements of interaction. Second, locations play a role by providing opportunities to access specific and locally bound resources (e.g. specialized workforce) and regional unequal distributed business opportunities (Sayer, 1991; Bathelt and Glückler, 2005). By analyzing US patent citations, Sonn and Storper (2008) confirm a positive effect of geographical proximity on the innovation output. However, geographical proximity is not the exclusive driver of collective innovation processes but also an enabler for realizing other forms of proximity (Boschma, 2005). The integration in networks is more important than mere geographical proximity (Breschi and Lissoni, 2003; Balconi, Breschi and Lissoni, 2004; Buchmann and Pyka, 2012a; Buchmann and Pyka, 2012b). Also, Boschma and ter Wal (2007) stress the importance of integration in global networks and value chains for innovative performance besides local network embeddedness.

## 7.6    Basic Sociological Concepts of Attachment Mechanisms

The question of whom to choose as a partner is not unique to firms which search a cooperation partner. The selection problem is likewise prevalent in other fields of life. For instance, human beings choose their partner preferably from their own race. As a general feature, in social networks there is a preference for the selection of partners which are similar (Newman, 2003). For this tendency, Lazarsfeld and Merton (1954) introduce the notion of *homophily*, defined as the formation of friendship between people of the same kind ("birds of a feather flock together"). Consequently, a determinant influencing the probability for the emergence of a tie between any two actors is their degree of similarity. This means that the propensity to cooperate is dependent on the actor characteristics and the embeddedness in a network as two actors can be similar in terms of (i) their structural position in a network or (ii) with regard to attributes such as size, reputation or resources including the knowledge-base. Therefore, two forms of homophily can be differentiated. Structural homophily (Podolny, 1994; Popielarz and McPherson, 1995) refers to the structural embeddedness in a network such as the degree centrality. Covariate related homophily (Van de Bunt and Groenewegen, 2007) relates to similar actor attributes. That is, the probability that a tie will be created between actors $i$ and $j$ is higher when they are similar in one or in a number of characteristics. The homophily effect is a reciprocity effect, i.e. when actor $i$ is similar to actor $j$, then actor $j$ must be similar to actor $i$.

In contrast to individual covariates, dyadic covariates are calculated for pairs of actors. A typical example would be all kinds of distance (proximity) measures between actors, such as geographic distance or technological distance. Dyadic covariates measure the extent to which the formation of a tie between two actors is more probable when the dyadic covariate is larger or respectively smaller. This kind of homophily is particularly interesting to study for the case of innovation networks where small technological distances indicate same understandings of problems (Giuliani and Bell, 2005).

## 7.7 Transitivity and other Triadic Effects

Even though I follow in my argumentation predominantly a knowledge-based view, I do not neglect that costs also play a role for selection processes. Consequently, there is a further factor which makes a selection process based on social criteria advantageous, namely costs which accrue for searching and evaluating the quality of a potential partner (Gould, 2002). The search for the ideal partner can be cumbersome, time consuming, cost intensive and requires capabilities of judgment. Especially small firms lack the necessary resources to conduct extensive search that covers all potential partners and produces reliable information for making a choice (Giuliani, 2010). When the level of uncertainty is high and hard facts about other firms are scarce or lacking, social signals play a strong role for reducing the circle of potential partners (Ibarra, 1993; Lazega et al., 2012). Firms are incentivized to create stable ties and relationships which enable mutual learning based on trust and cooperative behavior, reduce search costs and the risk of selecting the wrong partner (Powell, 1990). Garcia-Pont and Nohria (2002) study alliance dynamics among the 35 globally largest firms in the automotive industry. They find that the denser the network of the group is knit, the more their behaviors resemble and the higher is the probability that they select a partner from the group. Triadic structures, such as transitive triads, are a common pattern found in many networks. Transitivity is a structural effect which refers to the positioning of actors in a network. It describes a tendency of the partners $(i, j)$ of an actor $(k)$ to initiate a collaboration which leads to a closed triangle (Figure 13). The number of these triangles is expected to exceed the number of triadic structures in random networks (e.g. Davis, 1970; Holland and Leinhardt, 1971).

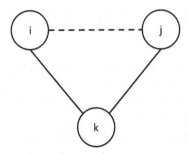

**Figure 13:** Triadic structure (Source: own illustration).

The formation of triads is an indication for the formation of dense interconnected cliques (Skvoretz and Willer, 1991). Gulati and Gargiulo (1999) find confirmation that common third-party ties between previously unconnected organizations increase their probability of initiating a collaboration. As firms operate in an environment of bounded rationality and imperfect information – also in terms of potential partners – they face the risk of opportunistic behavior (Gulati, 1995a). Acquiring reliable information about a potential partner is a difficult task, yet it is essential for the formation of alliances that serve both partners. Firms which share a common partner gain information about each other from their shared partner (Baker, 1990). Moreover, alliance partners have to cope with potential moral hazard threats due to the barely predictable behavior of the partner firm. For instance, one firm could try to free ride by making none or only very limited contributions to a common project while the other partner invests a lot more resources; or by using commonly developed knowledge in a way which damages the eligible interests of the partner (Gulati and Gargiulo, 1999). Whenever a firm is looking for a collaboration partner, existing links are most valuable and trustworthy sources of information about potential partners. For instance, if actor $j$ collaborates with actor $k$ and actor $i$ collaborates with actor $k$, then actor $k$ is a reliable source of information about the trustworthiness and reputation of actor $j$. This effect leads to closed triangles (the two-path $i \rightarrow k \rightarrow j$ is closed by the tie $i \rightarrow j$).

The formation of such closed triangles creates social spaces which are reinforced by shared beliefs and opinions that prevent actors from opportunistic behavior, it allows for the formation of trust and forges the exchange of tacit knowledge (Uzzi, 1997). The triadic structure may even lead to a reputation lock-in, i.e. there is a strong interest of the three actors that each partner behaves well to ensure a good reputation of the group. Selfish behavior will be reported to common partners and can be immediately penalized (Raub and Weesie, 1990; Burt and Knez, 1995). Consequently, triads tend to be more stable and durable compared to dyadic ties (Baum, Shipilov and Rowley, 2003).

## 7.8 Experience with Cooperation

A further factor I examine is a firm's experience with cooperation. For firms that have only little or even no experience at all with collaborative projects, entering a network is not a trivial task and various obstacles that hamper effective collaboration can be found in literature. According to Hennart (1988), Pisano (1989) and Parkhe (1993) the most prominent ones are: (i) lack of trust between the partners; (ii) unwillingness of sharing control and leadership in projects; (iii) overly high complexity of the project; (iv) uneven capabilities in learning; (v) confusions on the question about who is a partner and who is a competitor.

I suggest that a large record of collaborative activities signals a larger attractiveness as well as preference for further collaboration. Firms that are experienced cooperation partners find more easily additional partners (Ahuja, 2000b). This reflects that from outside it is rather difficult to scan a firm's valuable resources, in particular its knowledge-base. A firm which has been often involved in cooperative projects signals to be a valuable partner with a good reputation and established routines of collaboration. Furthermore, a long record of cooperation demonstrates that a firm made positive and obviously valuable experience with cooperative projects in the past. This will increase its willingness to initiate further collaborative projects in the future. Alliance or cooperation capabilities are specific and not transferable resources which enhance a firm's ability to identify a partner, initiate collaborations and manage a partnership successfully (e.g. Makadok, 2001). Experienced firms install dedicated collaboration management routines to coordinate the portfolio of different types of alliances (Kale, Dyer and Singh, 2002). Developing experience takes time since it forces a firm to adapt its internal routines (Powell, Koput and Smith-Doerr, 1996). However, it is worth the effort as it not only enables a firm to become effectively embedded in a formal innovation network but also paves the ground for informal collaboration (Pyka, 2000).

Clearly, history matters when decisions about future cooperation partners are made. Firms which have to make decisions about the configuration of their ego-network are confronted with a variety of potential hazards. In order to prevent the potential negative impacts, they make use of the information provided by already existing ties

(Powell and Smith-Doerr, 1994; Gulati, 1998; Gulati and Gargiulo, 1999). Decision makers use information about preceding alliances as a guideline for future partnerships as they provide reliable and timely information about the availability, capabilities and reliability of future partners (Gulati and Gargiulo, 1999). Availability heuristic may also play a role when managers decide about future alliances remembering past experiences with partners. The timely characteristic is particularly important in industries where time-to-market is a central determinant of competitiveness. Every new tie adds to the pool of information which constitutes the basis of decision making for the present and future. This information is derived from prior direct cooperation partners but also from indirectly connected firms and from the reputation that is linked to the positions of a potential partner in the pre-existing network (Gulati and Gargiulo, 1999). Prior ties between two firms increase the probability for establishing a trustworthy relation and reduce uncertainty with regard to future common projects (Podolny, 1994). Moreover, the regular interaction which is associated with ties between two firms may give rise to new ideas resulting in future projects that would be way more difficult to initiate if there was no pre-existing tie (Gulati and Gargiulo, 1999). Powell, Koput and Smith-Doerr (1996) find for the investigated network in the biotechnology industry that an estimated number of 15% of existing ties are terminated each year. Yet, the relationship often does not completely break up when the collaboration in a specific project comes to an (envisaged) end. This event may directly trigger a new project. For example, a common research project may be followed by a project of the same kind or by a common manufacturing project.

## 7.9    Conclusions

The evolution of interfirm innovation networks is driven by forces which have different points of application. Prior research on innovation networks suggests that there are drivers related to actor characteristics, to dyadic characteristics but also endogenous forces stemming from the network structure itself. Derived upon the concepts presented in this chapter (absorptive capacity, technological distance, knowledge-base modularity, geographical distance, transitivity and experience with cooperation), I test in chapter 9 the significance of hypothesized drivers for the particular case of a German automotive innovation network. Some of the presented effects are approximated with the information that is embodied in patent documents. In the following chapter, I explain which kind of patent information might be relevant and to what extent patent measures can be applied in the described context.

# 8.  Patenting

Patents serve as a rich source of information about inventions (and innovations). A large patent pool is a signal for a high level of competence and knowledge in one or more technology fields. Patents function therefore as an attractor for the search of a cooperation partner. A patent is a legal title granting its holder the exclusive right to make use of an invention for a limited area and time by preventing others from making, using or selling it without permission. In return, it forces the inventor to reveal the technical details of the invention. The aim of the patent regime is to solve the trade-off between *static* and *dynamic efficiency*. Static efficiency refers to the perpetuation of the incentives to innovate, while dynamic efficiency focuses on the socially desirable diffusion of new knowledge. The character of a legal right provides a patent with a (potential) economic value. From a technical point of view, a patent is the outcome of a successful patent application procedure and is granted by a patent office after it has been scrutinized for its validity. This scrutiny process is conducted by national or regional authorities who grant a patent or reject the application. The largest national patent offices are the USPTO (for the USA) and the JPO (for Japan). In principle, protection for an invention is only guaranteed in the country where a patent is granted and a patent has to be applied for in each country separately. However, for Europe the European Patent Office (EPO) facilitates the application process by granting a European patent. For validity in a specific country, the country still needs to be explicitly mentioned in the application document.[8] The system of national and regional patent offices is complemented by an international regime, the so called Patent Cooperation Treaty (PCT). With this scheme, pre-applications can be filed to a variety of national patent offices at the same time. A key advantage is that the costs are relatively low compared to the amount that applications to all the individual national offices would cost. The PCT-scheme is administrated by the World Intellectual Property Organization (WIPO) (Maraut et al., 2008).

A principal objective of the patent system is to sustain the incentives to innovate. When a firm allocates considerable resources to R&D, it expects a reasonable rate of return from the investment. Thus, there is a genuine interest on the side of the firms to protect the knowledge on which their inventions are based and to ideally ensure a monopoly for the exploitation. The existence of a link between (costly) innovative efforts and the opportunity to beneficially exploit the results is a central assumption in

---

[8] An agreement for a unitary EU patent (with the exception of Italy and Spain) was adopted during the European Council of 28 - 29 June 2012 (European Parliament, 2012).

many (innovation) economic models. Accordingly, a lack of appropriability regimes for profits derived from an invention is regarded as the main reason for underinvestment in R&D and low rates of innovation. In fact, technologies incorporate a mix of public and private good features (Arrow, 1962; Dosi, 1988). The latter allow firms to appropriate gains from their innovations and create incentives for further investments in R&D. Appropriation protects own novelties from being easily copied at the expense of a loss of own profits. The conditions of appropriation are industry and technology specific. Levin, Cohen and Mowery (1985) propose the following means of appropriation: patents, secrecy, lead time, costs and time required for duplication, learning-curve effects, superior sales and service effects. Dosi (1988) argues that due to partly tacit and partly private elements of technological knowledge, imitation is not a simple copy-paste process but requires creativity and resource employment, a task which is somewhat similar to the original inventory process and also costly.

By considering, for instance, the concepts of technological paradigms and trajectories, we may conclude that there are other reasons than a lack of appropriability responsible for differences in the rates of technological progress between firms and industries, such as limited technological opportunity spaces. For that reason, strengthening patent laws is an inadequate lever to increase the innovation rate in an economy and may even trigger *patent wars* which are foremost beneficial for the involved lawyers. (Too) strong patent laws may lead to a slowdown of technological progress if we presume a cumulative knowledge creation process in the sense that yesterday's achievements in the search for new solutions and protection of past inventions hamper researchers from implementing existing knowledge in their search routines. This conjecture holds particularly for very basic inventions that have a potentially broad field of application. Thus, there is no direct link between the degree of possible appropriability and a firm's efforts to innovate, and the different rates of technological progress cannot be traced back to differences in appropriability regimes (Dosi and Nelson, 2010). There is even a downside related to early and rigorous protection of inventions. As other firms are not allowed to apply the invention they cannot draw on this technology and develop and improve it even further. Thus, other firms cannot create additional learning opportunities for the firm which originally developed a technology.

## 8.1 Patents as a Proxy for Innovation

Entrepreneurial strategies to protect inventions and innovations differ across industries. While in the pharmaceutical and telecommunications industry patents are an important instrument and measure of innovation (Hagedoorn and Cloodt, 2003), this is less the case in other industries where time to market, learning curve advantages, secrecy or complementary assets and services are more important. Especially when firms introduce innovative (production) processes, it is often secrecy that serves as a means for the protection of new knowledge (Levin, Cohen and Mowery, 1985; Cohen, Nelson and Walsh, 2002). Moreover, some firms in an industry may not patent at all but still conduct R&D to understand what others are doing (Dosi, 1988). This is related to internal efforts which are necessary to establish an effective absorptive capacity (Cohen and Levinthal, 1990). In a direct way patents only represent the part of a firm's knowledge-base that can be codified, but this part is strongly correlated with measures reflecting the tacit component (Narin, Noma and Perry, 1987). The use of patent data for statistical analysis is seen as problematic if applied across industries (Hall, Jaffe and Trajtenberg, 2001). Consequently, one is advised to focus on one or few industries in the analysis or to analyze different industries separately from each other (Hagedoorn and Cloodt, 2003).

To measure the innovativeness of a firm, we are asked to identify some kind of proxy which is strongly correlated with the innovation output. Despite the described limitations, patents can be regarded as a valid measure for the output of the R&D process. Previous research shows that patents are indeed a valid indicator for the output, value and utility of inventions (Trajtenberg, 1990; Hall, Jaffe and Trajtenberg, 2005). They are a measure of invention which is externally validated during the application process at a patent office, and since this process is time consuming and costly, firms most probably launch the application process only for inventions which have some sort of potential economic or strategic value (Griliches, 1990). Moreover, there is a relatively large body of literature which employs patents not only as a proxy for invention but also for innovation (output) (e.g. Pavitt, 1985; Hagedoorn and Cloodt, 2003). By analyzing a large number of studies Hagedoorn and Cloodt (2003, p. 1368) conclude that "in large parts of the economics literature, raw patent counts are generally accepted as one of the most appropriate indicators that enable researchers to compare the inventive or innovative performance of companies in terms of new technologies, new processes and new products." Furthermore, Niosi (2005, p. 22) suggests: "Even if not all commercially useful novelties are patented, not all patents are exploited in the market, and the exploitation may occur in a place different from the one where the innovation took place, no other indicator is better suited to the study of innovation."


## 8.2   Application of Patent Data for Innovation Economic Analyses

In fact, the exploitation of patent data delivers rich information about the inventor, the applicant, the concerned fields of technology and knowledge flows (approximated by citations). For instance, from the addresses we can get geographic information which allows for testing hypotheses regarding the influence of geographic distance on cooperation. The regionalization of patent data is either based on the geographic coordinates of the inventor (a person) or of the applicant (firm, university, etc.). In the first case, we get to know the place where the inventor lives. Due to the fact that most inventors are not working in their own laboratory but are employed by a firm, university or research laboratory this is mostly also the place where the invention was made (as long as the inventor lives and works in the same region). In the second case, we get the information of the firm location. Some prudence is required as in some cases only the address of the headquarters is documented. Results from R&D cooperation leave sometimes traces on patent documents in the form of several named organizations on a single patent document. Also, inventor collaboration is a prolific mode of scientific work. Therefore, there is often more than just one inventor mentioned on a patent document. These kinds of information can be very helpful when we are looking for paper traces of R&D collaboration. In particular, the detection of large scale networks is challenging. One promising avenue is the use of patent data because they provide information on co-invention as well as citation patterns (Balconi, Breschi and Lissoni, 2004). Another possible application is demonstrated by Debackere, Luwel and Veugelers (1999) who use EPO patent data to proxy technological advantages. They link this data with a measure for comparative advantage based on trade data for the Belgium region of Flanders. By doing this, they search for a link between technological strength of a region and its economic strength.

To illustrate further kinds of studies which were conducted on the basis of patent data, I report some of the results:

- Gilsing et al. (2008) find that the innovation output increases with firm size but under-proportionately. Especially when it comes to the exploration of new technologies, small firms perform better. In addition, R&D intensity has a significantly positive effect on the innovation output and the age of the firm is negatively correlated (but non-significantly) with (exploratory) patents.

- When looking for explanations of innovative performance, Hagedoorn and Cloodt (2003) do not find a systematic disparity among the R&D inputs, patent counts, patent citations and new product announcements.

- De Rassenfosse and van Pottelsberghe de la Potterie (2009) study the relation between patent counts and R&D performance at the country level and find a strong correlation. Moreover, their results suggest that the propensity to patent and the research productivity are both relevant factors for explaining cross-country disparities in patent counts per scientist.

- Breschi, Lissoni and Malerba (2003) use patent data to test the hypothesis that technological relatedness is an important determinant of the diversification of a firm's knowledge-base. They find strong evidence for this hypothesis and conclude that, fueled by learning processes and the properties of knowledge, such as complementarity, firms predominantly focus their R&D efforts on technological fields which are related. Remarkably, even firms which have a very diversified technology portfolio mostly patent in fields which are knowledge-wise strongly related.

## 8.3   Building the Knowledge-Base from a Pool of Patents

Based on the previous discussion, I conclude that – as a proxy – patents can be regarded as the most elementary (discrete) building blocks of a firm's knowledge-base. This approach is in line with other studies reconstructing a firm's knowledge-base out of patents (see for instance Jaffe, 1989a; Ahuja and Katila, 2001; Fleming and Sorenson, 2001). For the purpose of the empirical study conducted for this dissertation, patent data are extracted from the OECD REGPAT database which contains patent (application) data that has been linked to geographic locations based on the addresses of inventors and applicants. All patent data included in the REGPAT database are taken from two primary sources, namely the EPO's Worldwide Statistical patent database (PATSTAT) and the Inventors and Applicants records from EPO patents extracted from Epoline web services. In the REGPAT June 2010 edition, the data is retrieved from the Patstat April 2010 edition and the complementary OECD patent database based on EPO's epoline@ database which covers publications up to June 2010. REGPAT covers patent applications filed to the EPO from 1977 to 2007 and partial data afterwards according to the priority data (OECD, June 2010).

For the regionalization of the patent data, 36 countries have been taken into account encompassing many OECD countries and a selection of European non-member countries. The REGPAT database includes applications to the European and US patent office (however incomplete) as well as applications filed under the umbrella of the Patent Cooperation Treaty (PCT). The link between inventors or applicants and a region was established by matching the postal codes or town names which are part of the address with regional units such as the NUTS3 regions. In general, researchers are interested in the date when the invention was made. This is best expressed by the priority year as it indicates when the applicant first filed for a patent. Other dates can also be found in the documents. The publication or grant date depend on the specific administrative procedures of the scrutinizing authority and can lie up to ten years after the invention (Maraut et al., 2008).

## 8.4    Conclusions

Despite a number of limitations, patents are the best publicly accessible proxy, both for inventions and innovations. In particular, for large studies there is no other comparable source which contains such rich information about inventors, applicants and technology fields. Consequently, I use patent data to map a firm's knowledge-base. Moreover, by using the information of the IPC classes which are documented on a patent, the structure of the knowledge-base can be represented as a knowledge network (cf. chapter 9.3).

# 9.    An Automotive Innovation Network

In this chapter, I first describe characteristics and challenges of German automotive suppliers and manufactures followed by an analysis of the structure of their collective knowledge-base, focusing on e-mobility technologies. Second, I introduce a model for the analysis of evolutionary change patterns of an interfirm innovation network (Pyka and Fagiolo, 2005; Pyka and Hanusch, 2006) which consists of a sample of German automotive firms. A stochastic actor-based model is applied to estimate parameters which reflect the impact of hypothesized effects. The elementary building blocks of the analyzed innovation network are nodes (firms) and ties (derived from collaborative R&D projects) representing interaction structures that serve as channels of implicit and explicit knowledge exchange. These basic network elements aggregate into a complex network structure which is embedded in a wider economic system. That is, an innovation network can be described as an integral part of the regional, national or sectoral innovation system.

Increasingly complex technologies in the automotive industry spur collaborative efforts of knowledge creation. Hardly any firm can maintain a leading role in competition by solely relying on isolated R&D endeavors. "[...] organizations can no longer hold mastery over all the emerging technologies which have the potential to impact on their products" (Birchall, Tovstiga and Chanaron, 2001, p. 86). Joint R&D projects, strategic alliances and other forms of collective innovation processes allow for the pooling of knowledge and competences (Teece, 1992). R&D cooperation opens channels to access critical resources. In contrast to the transaction cost approach (Coase, 1937) which focuses on cost minimization, Neo-Schumpeterian economists emphasize the importance of learning opportunities and the knowledge transfer processes in networks (Hanusch and Pyka, 2007a). Knowledge as the key resource for invention and innovation is scarce. It is hard to imitate, to transfer on markets and to substitute (Barney, 1991; Peteraf, 1993). Knowledge intense industries such as the automotive industry foster the general movement towards collaborative innovation (Powell et al., 2005; Pyka and Saviotti, 2005). "Collaborations are a useful vehicle for enhancing knowledge in critical areas of functioning where the requisite level of knowledge is lacking and cannot be developed within an acceptable timeframe or cost" (Madhok, 1997, p. 43).

Actor-based models for network evolution enable us to shed more light on the complex dynamics of continuously emerging and dissolving ties between firms (Ter Wal and Boschma, 2009) and are, thus, a useful instrument to learn about underlying micro mechanisms and to disentangle the driving factors of evolutionary change. For the analyzed innovation network, I consider actor characteristics (on the individual and dyad level) and social factors to be relevant drivers for network evolution. Based on the described theory (chapter 7), I test factors which are suggested to affect both, the propensity to cooperate as well as the preference for a certain type of partner. In particular, I suggest the following factors to be relevant: absorptive capacity, technological distance, the level of knowledge-base modularity, geographical distance, transitivity and experience with cooperation. In addition, I control for capacity effects such as the experience of a firm in the industry and the size of a firm, and for coordination costs of ties.

## 9.1    Industry Context: Cooperation in the Automotive Industry

A first step to understand network evolution is to understand its broader context. Single firms are part of industries and changes in industries feed back on individual firms (Brass et al. 2004). The growing importance of suppliers in design and production of components requires frequent interactions between suppliers and OEMs as well as among suppliers and among OEMs (Kotabe, Parente and Murray, 2007).

Due to the high requirements of data availability, the evolution of innovation networks was so far only studied in few industries and only few studies have focused on the mechanisms of evolutionary change over time: Ter Wal (2013) studies drivers of network evolution in the German biotechnology industry; Balland (2012) analyzes the global navigation satellite system industry (GNSS); Giuliani (2010) applies a dynamic network model in a study on a Chilean wine cluster and Balland, De Vaan and Boschma (2012) investigate the determinants of network evolution in the computer games industry. Traditional manufacturing industries, such as the automotive industry, have so far not been analyzed and the empirical understanding of innovation network evolution is still preliminary. Further research is needed to better grasp the role of industry specificities shaping innovation network evolution.

### 9.1.1   Industry Trends

The German car producers and suppliers are faced with a variety of challenges which force them to optimize their cost structure, and even more importantly, to search for innovative solutions with regard to their product portfolio and their organizational structure. To escape the selective pressure, new strategies are developed and implemented. Innovations allow firms to run-off a destructive price competition and to create unique selling propositions. However, in important future oriented technologies, such as hybrid engines, German firms are lagging behind (Dilk et al., 2008). Intensified innovation competitions as well as shortened product life cycles emerge as a race for innovation (Staiger, Gleich and Dilk, 2006). Furthermore, the automotive industry has undergone and is still undergoing a consolidation process which leads to (successful und unsuccessful) mergers and acquisitions. The mergers of Daimler-Benz and Chrysler and of Hyundai and Kia, the strategic alliance between Renault and Nissan as well as the takeover of Jaguar and Volvo by Ford and their resale to Tata Motors and Geely are just a few examples of this enduring process. Intensified competition, over-capacity and the catching-up of Asian firms challenge the old champions of the industry. During the 1980s already, Japanese car manufacturers increasingly formed strategic alliances on a global scale which provided them with a competitive advantage. Also, Rycroft and Kash (2004) identify globalization as an important driver for the proliferation of network structures in the automotive industry. Interestingly, there seems to be a co-evolutionary effect, that is, networks also push globalization dynamics: Technologies lead to both, changes in organizational structures, and the creation of more integrated markets and strongly rising trade volumes which feeds back on technological and organizational developments. Besides access to global value chains, firms need access to tacit and locally bound knowledge-bases of regional innovation systems to generate innovation for diversified and heterogeneous global markets. Innovation based on the ubiquity of codified knowledge (large data bases, ICT) complements locally sticky knowledge.

Increased complexity and new technologies, such as electro-mechanical integration, inter-connectedness of components and internet-based car solutions (Dilk et al., 2008) amplify the pressure to form alliances with partners operating at the cutting edge of technology. The complexity of cars rises sharply, making system integration an increasingly challenging task. On the other hand, technologies help to shorten time to market and increase flexibility by new design and engineering (rapid prototyping) tools, smart manufacturing facilities and collaboration. Stricter environmental regulation[9] requires solutions beyond the established design of the internal combustion engine fuelled with petrol. To face air pollution and climate change, the automotive industry finds itself increasingly under social and political pressure to produce more environmental friendly cars. German producers are particularly affected by regularity

---

[9] For instance, EU Regulation 443/2009 forces car producers by 2020 to reduce $CO_2$ emissions of their product portfolio to a level which does not exceed the threshold of 95g $CO_2$/km.

hurdles since their cars are known for being high comfort which went in the past hand in hand with heavy weight and high emissions. The dominant design of the internal combustion engine, as the heart of the power train, is increasingly challenged by new and supposedly more efficient technologies. With the established design being challenged also the "masters" of this design, the incumbent car manufacturers and their suppliers, are threatened by new firms appearing now on the playing field. New technologies leverage the possibilities and lower market entrance barriers for innovative firms. New solutions lead to an erosion of the value of incumbent knowledge-bases if they are not "refilled" with new knowledge. A modular structure of the knowledge-base helps to adopt such new technologies (chapter 7.4). With electric cars, completely new components are needed in a number of fields for which the incumbents not necessarily have the required expertise. This concerns, i.a. the power train (e.g. electric engine, gear box), the battery, brakes, electronic control units, climatisation, light bodies and the chassis. Taken together, the described changes open a window of opportunity for new players from inside and outside the industry to enter the market.

The automotive industry is characterized as a scale driven but likewise knowledge-intensive industry. The advent of new technologies in tandem with a high level of uncertainty concerning the power train, assistant systems as well information and communication technologies, triggered increasing R&D efforts during the last years. In a discussion with an automotive expert of a large consulting firm I asked the question if joint R&D projects are seen in the industry as a means to cope with technological uncertainty. The expert made the point that "a high level of uncertainty with regard to future dominant technologies is a strong driver for cooperation. None of the many small suppliers has a clue of what will become a standard. The network serves as a laboratory to do 'experiments' without taking too much risk" (Roland Berger, 2010). A study of Deloitte Consulting (2009) identifies a current phase of industry convergence, i.e. in the current phase of the life cycle new players enter the field and cooperation is not anymore an intra-industry phenomenon but takes place on an inter-industry scale. For instance, battery producers cooperate with climatisation experts in cooling technologies. A particular active field of current convergence processes is research and development. As a kind of anecdotal evidence, this gets confirmed by (unsystematically) collected information on collaboration projects announced in the German (business) newspapers Financial Times Deutschland, Handelsblatt and Frankfurter Allgemeine Zeitung between September 2010 and August 2012 as presented in Table 2. The collected information indicates that the OMEs cooperate among each other to develop technologies that lie beyond the current paradigm, but also with suppliers, such as SGL Carbon, that have, for instance very, specific competences in the field of light materials.

**Table 2:** Collection of collaborative projects.

| Date | Source | Firm (i) | Firm (j) | Motivation |
|---|---|---|---|---|
| 17.9.2010 | Financial Times Deutschland | Daimler | Toyota | Access to Toyota's hybrid technology |
| | | Daimler | BMW | Development of hybrid technology |
| | | Renault | Nissan | Development of environmentally friendly power trains |
| | | Daimler | Evonik | Development of batteries for electric cars |
| 13.7.2011 | Handelsblatt | Daimler | Bosch | Common factory for electric engines (joint venture) |
| 29.7.2011 | Handelsblatt | Daimler | Renault | Renault Twingo (electric) receives battery from Daimler; Common development of Twingo and Smart; Renault provides platform of Kangoo; Common development of engines |
| 24.8.2011 | Handelsblatt | Bosch BASF Thyssen-Krupp | | Production of high capacity batteries with lithium-ion technology |
| 29.8.2011 | Handelsblatt | Opel | 30 suppliers of renewable energy | Renewable energy for electric cars |
| 1.9.2011 | Handelsblatt | Siemens | Volvo | Electric cars |
| | | GM | LG | Development of electric cars |
| 2.9.2011 | Handelsblatt | Daimler | RWE | E-mobility |
| | | BMW | SGL Carbon | Common factory for carbon fibers (joint venture) |
| 21.9.2011 | Financial Times Deutschland | GM | SAIC | Common development and manufacturing of electric cars; access to Chinese market |

|  |  | Volkswagen | SAIC | Access to Chinese market |
|---|---|---|---|---|
| 29.2.2012 | Frankfurter Allgemeine Zeitung | GM | Peugeot | Equity alliance (GM buys shares of PSA) for common product development of Opel and Peugeot |
|  |  | Peugeot | Mini | Peugeot delivers diesel engines |
| 5.3.2012 | Frankfurter Allgemeine Zeitung | Jaguar | Chery | Common car manufacturing in China |
| 15.3.2012 | Frankfurter Allgemeine Zeitung | Volkswagen | MAN | Volkswagen delivers power trains; MAN complements portfolio with small trucks |
| 16.4.2012 | Frankfurter Allgemeine Zeitung | Ford | Dow | Development of light material (carbon fiber) |
|  |  | General Motors | Teijin | Development of light material (carbon fiber) |
|  |  | Volkswagen | SGL Carbon | Development of light material (carbon fiber) (VW buys shares) |
|  |  | BMW | SGL Carbon | Development of light material (carbon fiber) (BMW buys shares) |
|  |  | Daimler | Toray | Development of light material (carbon fiber) (Joint Venture) |
|  |  | Daimler | BASF | Development of Smart Forvision (concept car) |
| 24.8.2012 | Handelsblatt | Daimler | Renault-Nissan (Infiniti) | Cooperation for development and production; Cost reductions |

In a stylized way, the automotive industry stands in a life cycle concept between the "alliance" phase and the "restructuration" phase (Figure 14). If we consider the power train as the core module of a car, the dominant design of the internal combustion engine became established at the end of a pioneer phase. In the following phases, changes in technology (in particular with regard to the power train) where more or less incremental only. The paradigm shift towards new concepts of the power train, which we now observe, requires more radical new knowledge and thus a renewed industry knowledge-base (cf. chapter 9.3). This pushes firms to enter R&D alliances which are characteristic for the current phase. Consequently, the industry is leaving a rather exploitative phase and is entering a more explorative phase. A new phase of the industry life cycle takes off which is characterized by the integration and development of new knowledge. It strengthens the explorative side and requires strong absorptive capacities to acquire and process external knowledge which might even stem from external industries.

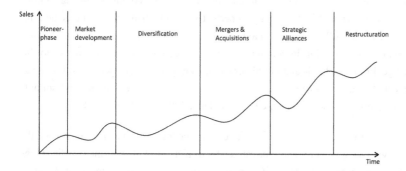

**Figure 14:** Industry lifecycle (Source: own illustration based on Deloitte Consulting, 2009).

This explorative side of the industry finds its counterpart by the attempt to exploit the existing knowledge-base in the most efficient way. In the automotive industry, both, product and process innovation play important roles. Moreover, process innovation (e.g. production technologies yielding economies of scale) plays an important role for a continuous growth in productivity (Van Biesebroeck, 2003). The ongoing consolidation of the industry and the high correlation of sales with the general business cycle forces producers to build flexible production facilities and cultivate close relations with their suppliers. Sophisticated production technologies and a high level of flexibility in combination with modular product architecture[10] create advantages in competition by allowing the realization of economies of scale and flexibility at the same time.

### 9.1.2   The Importance of the OEM-Supplier Network

The basic structure of the automotive industry is characterized by few OEMs (Original Equipment Manufacturers) and a large number of suppliers grouped in different tiers.

---

[10] See for instance the MQB (Modularer Querbaukasten) architecture of Volkswagen.

The latter group encompasses component manufactures (often SMEs) as well as big multinational enterprises (e.g. Bosch, ZF) which assemble entire systems that are just in time supplied at the assembly lines of the OEMs. During the last decade, more and more value creation (including R&D), and with it relevant knowledge, was shifted from the OEMs to specialized suppliers (Chanaron and Rennard, 2007). This organizational shift together with increased complexity of parts and systems created new coordination and transaction problems along the value chain. Electronic systems linking various units of a car need to communicate with a common language and be able to interact without interference in a perfectly reliable manner. Taken together, this means that the different parts have to be developed within a comprehensive framework. Collaboration in common research projects is seen as the answer to this challenge. From single bi- and multilateral collaboration projects, networks develop as a strategic instrument for the long run, i.e. firms envisage stable relations and not only single common  projects (Staiger, Gleich and Dilk, 2006; Dilk et al., 2008).

Due to the network character of the entire car production and development process, the costs as well as the quality and innovativeness of a car are linked to the supplier-OEM network. For instance, Dyer (1996) finds that firms which create specialized supply networks are more successful than competitors. The quality of collaborative component development is related to at least three fields of producer-OEM interactions: the approach of problem solving, communication pattern and the size and quality of the knowledge-base. Hence, for the analysis and explanation of success or failure of automotive clusters, the relational view of the network is a most promising approach. In highly integrated production systems, such as we find them in the automotive industry, the competitiveness of the system integrators (OEMs) is highly dependent on supplier capability and on how well the involved firms manage the division of labor (Takeishi, 2001). A large number of very "specialized suppliers" do product innovation primarily based on informal R&D, tacit knowledge and in close relation with their customers. Moreover, advanced production processes imply that firms master complex systems (manufacturing of complex products) which entail high R&D investments. At the same time, economies of scale and mass production are important. A classic example for a dense and strongly knit supplier-OEM network is the *Toyota supplier association*. Due to highly developed knowledge transfer routines, membership is positively correlated with high productivity performance. To facilitate the transfer of valuable tacit knowledge and to effectively disseminate the Toyota production system knowledge, a key objective from the beginning was to form strong ties (Dyer and Nobeoka, 2000). The shift of value creation and R&D to suppliers requires that the OMEs actively manage the network. Takeishi (2001, p. 419) summarizes the necessary efforts for the OEMs with the words: "Ask not what your suppliers can do for you; ask what you can do with your suppliers." Thus, there are three main goals for the installation of automotive networks (Dilk et al., 2008):

- Access to internally not available knowledge and technologies
- Contact with clients and downstream market participants
- Long-term bonding of suppliers and clients

## 9.2 Selection of Relevant Network Actors

For an empirical analysis of the evolution of an automotive network, the first step is to select the firms which are considered to be (potentially) part of the network. This opens the discussion about the boundaries of the network which is a topic for its own in the literature on social network analysis (e.g. Laumann, Marsden and Prensky, 1992) and will hence not be discussed extensively in this dissertation. The aim at this point of the dissertation is to study a publicly funded innovation network in the German automotive industry. While it is relatively easy to filter German firms by their location (address), the approach for capturing firms of a specific industry, that is a sample of firms selected according to some criteria, is more contentious. In light of the convergence process, delineated in the previous subchapter, which brings new players in the game, the reliance on standard classifications such as NACE (Nomenclature statistique des activités économiques dans la Communauté européenne) can easily lead to samples that are lacking important actors. Since the general line of argumentation in this dissertation is led by a knowledge-based view of the firm, the composition of the sample (which must not be confused with a probability or random sample) is based on the character of firms' patent portfolios. To identify relevant firms, I scanned in a first step the patent portfolios (OECD, June 2010) REGPAT database (unadjusted) of the largest German automotive OEMs and the largest suppliers. A ranking of the IPC classes (3-digit) by their relative frequency of occurrence shows that the class "B60" (vehicles in general) is the dominant patent class in the OEMs' portfolios and strong in the suppliers' portfolios (patent priority years 1977-2008). Moreover, the top 9 classes account for 68% - 91% of all the patents that belong to the selected firms.

**Table 3:** Distribution of relevant IPC classes.

| IPC (3-digit) | Volkswagen | Porsche | Daimler, Mercedes | BMW | Bosch | Continental | ZF |
|---|---|---|---|---|---|---|---|
| B60 | 30% | 31% | 26% | 27% | 16% | 38% | 19% |
| F02 | 17% | 14% | 13% | 14% | 26% | 11% | 1% |
| F01 | 10% | 12% | 7% | 10% | 2% | 1% | 1% |
| F16 | 10% | 19% | 7% | 10% | 5% | 8% | 57% |
| G01 | 5% | 2% | 5% | 4% | 11% | 8% | 2% |
| B62 | 5% | 10% | 7% | 8% | 0% | 2% | 9% |
| H01 | 2% | 0% | 5% | 3% | 6% | 3% | 1% |
| B29 | 1% | 1% | 1% | 1% | 0% | 6% | 0% |
| F04 | 0% | 0% | 1% | 1% | 2% | 3% | 2% |
| Sum | 79% | 90% | 72% | 78% | 68% | 81% | 91% |

Source: own calculations.

Both, the widespread use of patents to protect inventions and the high R&D intensity of the automotive industry are indications for the importance of patents in the automotive industry. Accordingly, I picked in a next step all firms from the OECD (June 2010) REGPAT database which filed at least one patent application in the class "B60" within the period 1998 to 2007, and I neglected non-patenting firms. Firms which do not hold patents are most likely unimportant actors in the industry from a technological point of view (Yayavaram and Ahuja, 2008). In addition, I discarded those firms which were exclusively operating in the market for commercial vehicles or car accessory kits based on information from companies' websites. Hence, I excluded all firms which were not directly related to the production of passenger cars. I also excluded firms which have not been involved in at least one collaborative research project during the observation period. The design of the sample according to this "recipe" resulted in a selection of 153 firms belonging to the network sample (Table A. 2).

The analyzed networks are reconstructed based on information retrieved from the German "Förderkatalog" (subsidies catalogue). This is a database which contains rich information about all kinds of research projects funded by the federal government. The database is publicly accessible via the website *www.foerderkatalog.de*.[11] Only those firms were eventually picked for the analysis which participated during the period

---

[11] On the European level a similar database is available covering the projects of the European framework programs. For the US, there is the National Cooperative Research Act-Research Joint Venture (NCRA-RJV) database of US-based research joint ventures research, a longitudinal database of strategic technical alliances. The NCRA-RJV database contains all RJVs registered with the US Department of Justice under the National Cooperative Research Act of 1984 and its amendments in 1993 (Vonortas, 2009).

1998-2007 at least once in a funded project. In order to model network evolution, the first step is to collect observations. The question is: What is an observation in a study on complete networks? We will see that for this kind of analysis the entire network is considered an observation. For the network reconstruction, the following assumption is made: A tie emerges between two actors $i$ and $j$ if they participated in the same project (see Broekel and Graf (2010) for this approach with the "Förderkatalog" database). Despite the fact that the database contains rich information about subsidized cooperative research projects, it has thus far rarely been used (Broekel and Graf, 2010). This is surprising because compared to patent data, information on joint projects documents research activities of firms in an earlier stage of the innovation process. In this phase R&D subsidies are used as a policy tool not only to incentivise and to channel R&D investments into new technologies, but also to support the exploration of knowledge synergies and mutual knowledge generation.

R&D subsidies for collaborative research have become a common tool for innovation policy makers for a number of reasons: First, due to the sheer scale and broadness of some projects they cannot be operated by single firms. Second, knowledge transfer from public to private organizations shall be fostered by the participation of universities and other public research facilities such as Max Planck and Fraunhofer Institutes. Third, collective learning processes shall be fostered (Broekel and Graf, 2010). In German innovation policy, elements of collaborative research in the design of innovation policies gained considerably in importance since the 1990s and include cooperation among firms as well as cooperation between firms and public research institutes. The projects listed in the "Förderkatalog" contribute to knowledge transfer and collective learning (Broekel and Graf, 2010): The participants have to sign agreements explicitly stipulating that generated knowledge within the project is freely shared among participants. They even have to grant free access to their know-how and IPRs within the scope of the projects. Furthermore, they commit to collaborate actively with the aim to find new solutions (BMBF, 2008). In the year 2002, almost 70% of direct subsidies in the field of mobility and traffic were assigned to collaborative projects (Czarnitzki et al., 2003). Moreover, theses funding schemes have a proven economic relevance: Czarnitzki, Ebersberger and Fier (2007) find that R&D subsidies influence collaborating and patenting activities. Fornahl, Broekel and Boschma (2011) find for the German biotech industry that R&D subsidies for collaborative research lead to increased patent output.

To reconstruct (two-mode) networks from project data, at least the following information is needed: name of the project, starting and end date as well as the names of the participating organizations. In addition, we can find information about the grant, the location of the receiving/executing organization and a classification number which divides funded technologies into different classes like biotechnology, energy etc. The title of the project is important to separate cooperative projects ("Verbundprojekt" or "Verbundvorhaben") from non-cooperative projects in which single organizations are funded. If two firms participate in the same project, this affiliation leads to joint activities, interactions and exchanges. "Thus, a two-mode network often goes together with interactions that can be described by one-mode networks" (Snijders, Lomi and Torló, 2013, p. 265).

As an example, say project "*I*" is a cooperative research project in which the four firms *A, B, C* and *D* collaborate. The participating firms are only considered if they fulfill the before mentioned criterion: They must have applied for at least one patent to which the IPC class "B60" was assigned. If, for instance, firm *D* does not fulfill the criterion, it is not considered a relevant actor for the analysis. Accordingly, firm *D* is not part of the network but the other three firms *A, B* and *C* are connected with each other forming a closed triad (Figure 15). In this way, all projects of the database which involve at least one of the actors from the sample are analyzed and the network for the entire sample gets reconstructed.

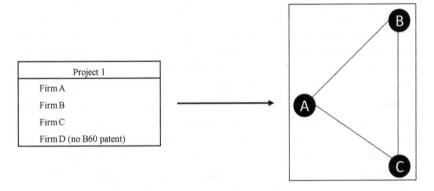

**Figure 15:** Network mapping of project partnership (Source: own illustration).

Sources containing information about interfirm networks, in particular for longitudinal network studies, are scarce. The application of the "Förderkatalog" database to reconstruct networks can be regarded as a complementary source to more established and "exploited" sources such as patent data or publication data. Still, patent data provide valuable information on the knowledge-base of the analyzed firms.

## 9.3 Analysis of the Automotive Sample Knowledge-Base

Patent documents are classified by one or more classification codes of the International Patent Classification (IPC) scheme (cf. chapter 8). One of the basic methods to analyze a knowledge-base which consists out of patents is to simple count the technology classes (IPC) that are assigned to a firm's patents. However, Engelsman and Van Raan (1991), Breschi, Lissoni and Malerba (2003) as well as Saviotti (2004) identify characteristics of knowledge structures which imply a different kind of analysis. Their analysis is based on the relatedness between technology fields which is approximated by the co-occurrence of IPC codes assigned to patents. The following assumption is made: The frequency of co-occurrence of IPC codes on the same patent is a proxy for the strength of technology and knowledge relationship. A possible objection towards the application of IPC codes to approximate the relatedness of knowledge refers to the fact that they are assigned by patent examiners and do allegedly not necessarily reflect the firm's perception of relatedness between technology fields. However, as patent examiners classify all technical aspects of the invention, the assignation of multiple codes reflects in a more impartial way technological relatedness. Before a code gets assigned, other patent documents are scanned and if a technological feature is found in another document, the respective code gets assigned.

In this subchapter, I explore the knowledge-base of the selected automotive firms and analyze some interesting characteristics such as the degree distribution, the tendency towards a small-world network, the development of the centrality of e-mobility patents and the influence of ties strength on overlapping ego-networks. In particular, I investigate to what extent the emergence of e-mobility technologies is reflected in changes of the knowledge-base structure of the firm sample. Moreover, I test if a tendency for overlapping ego-networks of IPC classes correlates with the tie strength.

### 9.3.1 The Knowledge-Base as a Network

The correlational and interpretative structure of knowledge makes it possible to analyze a knowledge-base as a network (Loasby, 2001; Saviotti, 2004). Elements of a knowledge-base are typically not independent from each other but there is some kind of relatedness between them. The aim must be to understand the relational structure between knowledge elements and its implications (Saviotti, 2004). For the analysis of the knowledge-base of the sample of 153 selected firms (chapter 9.2), the network is reconstructed as a dichotomized structure, i.e. it is checked if a tie between two IPC sub-classes $(i,j)$ is present $(X_{i,j} = 1)$ or absent $(X_{i,j} = 0)$. Present means that two IPC sub-classes (4-digits) co-occur on the same patent. Implicitly the assumption is made that co-occurrence means that there is a connection between the two elements. I look at the patent data through a five year moving window, i.e. five networks of consecutive time windows are reconstructed with the first window encompassing the patents first applied for between the years 1998 and 2002 (priority date).

**Table 4:** Industry knowledge network characteristics.

|  | 1998-2002 | 1999-2003 | 2000-2004 | 2001-2005 | 2002-2006 |
|---|---|---|---|---|---|
| Number of nodes | 473 | 467 | 454 | 459 | 458 |
| Number of ties | 6002 | 5958 | 5817 | 5312 | 4902 |
| Density | 0.054 | 0.054 | 0.057 | 0.051 | 0.047 |

Source: own calculations.

Table 4 indicates that the number of nodes, which is the number of occurring IPC sub-classes, remains relatively stable with the lowest value in the period 2000-2004 being however only 4% lower than the highest value in the initial period. In contrast, the number of ties is constantly decreasing with the lowest value in 2002-2006 being 18% smaller than the highest value in the beginning. The density is relatively stable in the first periods but falls back in the last period to a value of 4.7 %. A decreasing density measure can be interpreted as a transition from an exploitative phase in the life cycle to a more explorative phase. Exploration requires the recombination of old knowledge with (for the industry) new knowledge and thus new nodes entering the network. The formation of new ties is not expected to occur simultaneously with the occurrence of new ties. Consequently, the density measure falls (Saviotti, 2009).

We often find in networks degree distributions that are not homogenous across the nodes. For instance, Saviotti (2009) shows for a sample of pharmaceutical firms that the tie distribution and strength are highly heterogeneous. If we have a look at the (dichotomized) degree distributions in the sample (Figure 16), we identify two peculiarities: First, the number of low degrees is high compared to a normal distribution. Second, the right tails of the curves are relatively fat, especially for the time windows 2000-2004 and 2001-2005. This indicates that there are relatively many technology classes with high degrees compared to a normal distribution. This first graphical analysis gets confirmed in Figure 17 which shows quantile - quantile plots (Q-Q-plots) of the network degree distributions against (theoretical) normal distributions and a simulated normal distribution against a normal distribution (bottom right) as a control. For the degree centralities to be normally distributed the points need to be positioned on the middle line. This is only the case for a simulated normal distribution in the lower right but not for the empirical network degree centralities. The two lines which are parallel to the diagonal line represent the confidence bounds at a significance level of 5%. Obviously, the vast majority of points lies outside the boundaries. In particular, the graphics show that the distribution is right skewed. Besides this visual test for normal distribution, I ran a Shapiro-Wilk normality test (Shapiro and Wilk, 1965). The null hypothesis states that a sample (vector of metric values) is derived from a normally distributed population. From the test we get a p-value which has to be compared to a chosen alpha level. If it is smaller than the alpha level, then the null hypothesis is rejected (i.e. we can conclude that the data do not come from a normally distributed population). For the tested degree distribution the p-values are all very small (i.e. smaller than 0.05). Thus, the conclusion can be drawn that the degree distributions of the tested networks are not normally distributed.

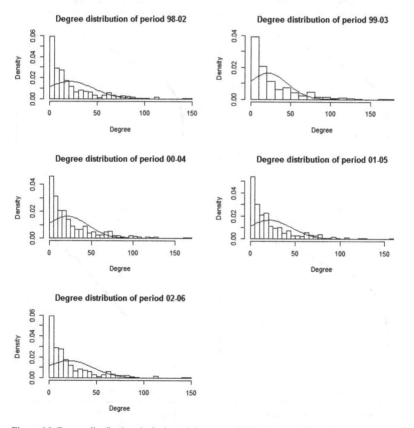

**Figure 16:** Degree distributions in the knowledge network (Source: own illustration).

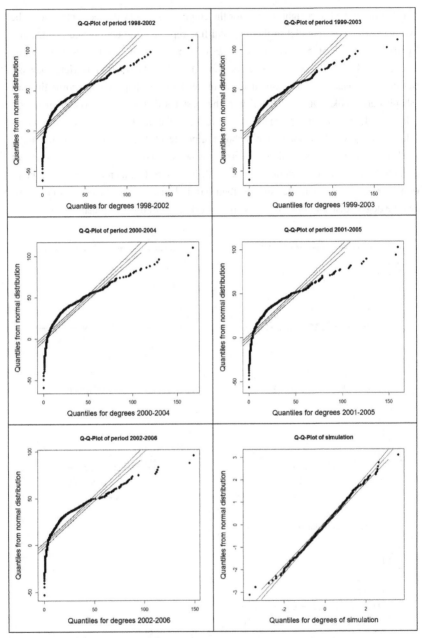

**Figure 17:** Q-Q-Plot degree distribution (Source: own illustration).

Table A. 1 illustrates the full list of degree centrality measures for each IPC sub-class and in particular its relative position among all sub-classes in the sample separated by the five consecutive windows of analysis. I suggest that important knowledge fields, represented by IPC sub-classes, are not only reflected by increasing numbers of patents but also by an increasing relative importance reflected by growing rates of degree centrality. The empirical data show that for all five observation periods, the top two sub-classes remain the same with the sub-class B60R (vehicles, vehicle fittings or vehicle parts) on the first position for the window 1998-2002 and 2002-2006, and the sub-class H01L (semiconductor devices; electric solid state devices) being number one during the three windows in between.

An important topic in the automotive industry remains the current shift towards e-mobility and hybrid powertrains. German automotive firms have in the past been frequently criticized because of the alleged little efforts spent in the field of e-mobility and particular in hybrid technologies, whereas, Japanese manufacturers (in particular the Toyota group) forcefully advertized their hybrid cars. The question arises, if this criticism finds support by an analysis of the sample knowledge-base. Karl and Jäger (2011) identify IPC sub-classes that are related to e-mobility and hybrid technologies, such as batteries, electric engines, control units etc. (Table 5).

**Table 5:** Important IPC sub-classes for e-mobility.

| Number | IPC sub-class | Technology |
|---|---|---|
| 1 | H01M | Battery |
| 2 | B60L | Propulsion |
| 3 | B60K | Propulsion unit |
| 4 | H02J | Supplying, distributing and storing of electric power and energy |
| 5 | H02K | Dynamo-electric machines |
| 6 | G01R | Measuring |
| 7 | B60H | Climate control |
| 8 | B60W | Control systems for hybrid vehicles |
| 9 | B60R | Vehicle fittings |
| 10 | H02P | Control or regulation of electric motor |
| 11 | B60T | Vehicle brake control systems |
| 12 | H01R | Cables |
| 13 | H02M | Apparatus for conversion (ac-dc etc.) |
| 14 | F16H | Gearing |
| 15 | B62D | Motor vehicles; trailers |
| 16 | H01L | Semiconductor devices |
| 17 | F02D | Controlling combustion engines |
| 18 | H02G | Installation of electric cables or line |
| 19 | H05K | Cooling |

| 20 | H02H | Emergency protective circuit arrangements |
|----|------|-------------------------------------------|
| 21 | H01B | Cables; Conductors; Insulators |
| 22 | B60Q | Signaling or lighting devices |

Source: own illustration based on Karl and Jäger (2011).

Moreover, Karl and Jäger (2011) find for the years 2000-2006 a rather constant number of e-mobility (including hybrid technology) patent applications of German automotive firms and strongly rising application activities only in the years afterwards. The same result is found by a study of Stahlecker, Lay and Zanker (2010). If we have a look at the degree centralities of the respective e-mobility sub-classes (Table 6), we find confirmative results. The degree centrality measure informs about the frequency the respective knowledge is used in conjunction with other knowledge fields. Overall, knowledge covering e-mobility classes rank high relative to other knowledge fields. However, the dynamic is low, only the sub-classes B60T (braking) and H02P (electronic engine control) constantly improve their position.

**Table 6:** Degree centralities of e-mobility sub-classes (left: degree centrality measure, right: position).

| B60R | | H01L | | H05K | | B60K | | H01M | |
|------|---|------|---|------|---|------|---|------|---|
| 178 | 1 | 174 | 2 | 122 | 5 | 119 | 6 | 115 | 7 |
| 166 | 2 | 179 | 1 | 118 | 7 | 121 | 5 | 120 | 6 |
| 163 | 2 | 168 | 1 | 110 | 8 | 130 | 3 | 120 | 6 |
| 158 | 2 | 160 | 1 | 109 | 6 | 123 | 4 | 107 | 8 |
| 149 | 1 | 145 | 2 | 111 | 6 | 113 | 4 | 83 | 12 |
| B62D | | F16H | | F02D | | H02K | | B60T | |
| 112 | 9 | 109 | 10 | 97 | 13 | 94 | 15 | 93 | 16 |
| 112 | 9 | 103 | 10 | 99 | 12 | 97 | 13 | 94 | 15 |
| 116 | 7 | 96 | 15 | 98 | 13 | 101 | 12 | 98 | 13 |
| 108 | 7 | 87 | 15 | 92 | 12 | 91 | 13 | 93 | 11 |
| 94 | 7 | 87 | 9 | 82 | 13 | 84 | 11 | 88 | 8 |
| H01R | | G01R | | B60H | | B60Q | | B60W | |
| 80 | 24 | 76 | 28 | 70 | 33 | 70 | 33 | 56 | 43 |
| 82 | 18 | 82 | 18 | 73 | 26 | 71 | 28 | 54 | 42 |
| 80 | 21 | 81 | 20 | 72 | 27 | 67 | 32 | 51 | 46 |
| 76 | 19 | 79 | 16 | 72 | 21 | 63 | 28 | 44 | 43 |
| 64 | 26 | 77 | 16 | 67 | 24 | 62 | 28 | 39 | 45 |
| B60L | | H02J | | H02P | | H01B | | H02H | |
| 50 | 49 | 49 | 50 | 48 | 51 | 44 | 55 | 43 | 56 |
| 53 | 43 | 46 | 50 | 46 | 50 | 43 | 53 | 41 | 55 |
| 52 | 45 | 43 | 53 | 46 | 50 | 48 | 48 | 40 | 56 |
| 44 | 43 | 41 | 46 | 47 | 40 | 45 | 42 | 44 | 43 |
| 35 | 49 | 42 | 42 | 45 | 39 | 39 | 45 | 42 | 42 |

| H02M | | H02G | |
|------|------|------|------|
| 37 | 59 | 27 | 68 |
| 35 | 60 | 26 | 68 |
| 39 | 57 | 26 | 70 |
| 35 | 52 | 28 | 59 |
| 33 | 51 | 26 | 58 |

Source: own calculations.

Besides the degree centrality, networks contain an additional piece of information, namely the tie strength. The analysis of the sample knowledge-base becomes enriched by taking the tie strength into account. Its analysis in a network view has the potential to extend our understanding of knowledge-base structures. Figure 18 shows that the order of the top ten nodes in terms of degree centrality is relatively stable. Same color means that the nodes belong to the same IPC class (3-digit) while the size of the nodes indicates their degree centrality. In period 2001-2005 the sub-class G06F (electric digital data processing) enters the top ten and in period 2002-2006 the sub-class B60T (vehicle brake control systems). Across all observations the strongest tie among the top ten nodes is the one between B60R (vehicle fittings) and the general sub-class B62D (motor vehicles; trailers). These fields of knowledge are most strongly related. The tie between H01L (semiconductor devices) and F02M (supplying combustion engines with combustible mixtures) is constantly getting weaker. Also the tie between H01L and H05K (cooling) is getting weaker. In contrast, the tie between H01L and G06F (electric digital data processing) is getting stronger in the last observation period.

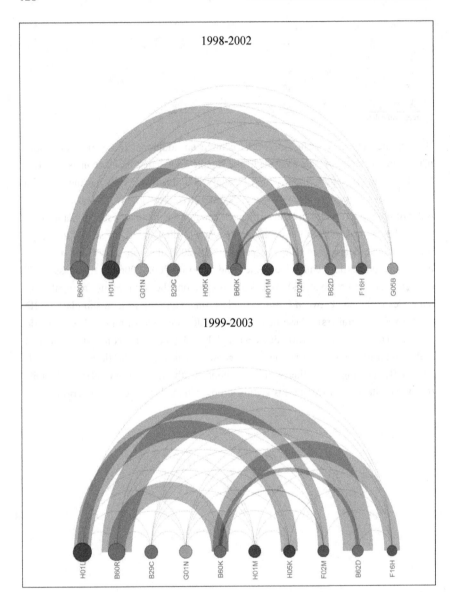

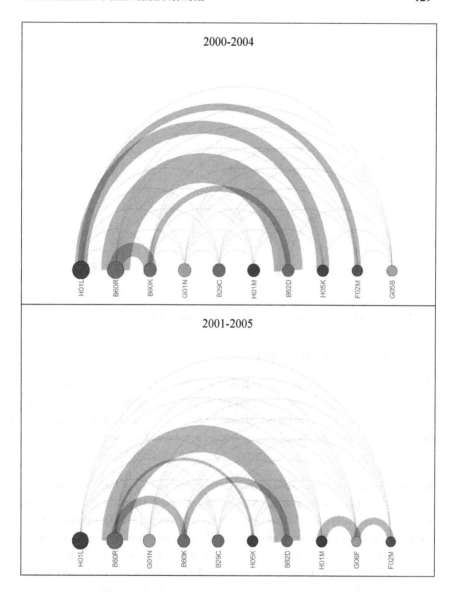

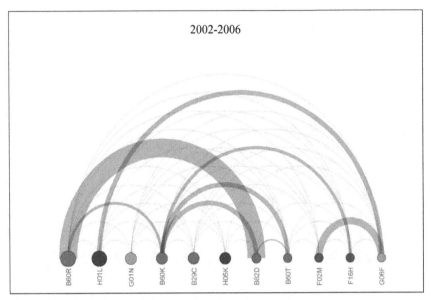

**Figure 18:** Tie strength between IPC-classes (Source: own illustration).

## 9.3.2  Test for Small-World Properties

As illustrated in chapter 5.4, a widespread feature of networks is small-world characteristics with a short average path length and a pronounced tendency for the formation of densely interconnected cliques. To find out if this characteristic can also be found in the analyzed knowledge-network, I test the industry knowledge-bases in five consecutive periods for their propensity to exhibit small-world properties.

Small-world networks are characterized by two features: (i) a high level of local clustering and (ii) a short average path length between network actors. To test for small-world characteristics, I draw on the Watts and Strogatz (1998) approach which compares the observed network path length ($PL$) and clustering coefficient ($CC$) with the respective properties of a random network with the same size and same number of ties. To quantify the comparison, the small-world quotient ($Q$) is applied. It is defined as the ratio of the (global) clustering coefficient ($CC/CCr$) divided by the ratio of the average path length ($PL/PLr$). The extension $r$ refers to the respective value calculated from a random network.

**Table 7:** Small-world test.

| | Knowledge Network 1998-2002 | Knowledge Network 1999-2003 | Knowledge Network 2000-2004 | Knowledge Network 2001-2005 | Knowledge Network 2002-2006 |
|---|---|---|---|---|---|
| Nodes (largest component) | 459 (97%) | 457 (98%) | 447 (99%) | 447 (97%) | 447 (98%) |
| Ties | 5998 | 5955 | 5814 | 5309 | 4901 |
| CC | 0.33 | 0.33 | 0.32 | 0.32 | 0.31 |
| PL | 2.41 | 2.42 | 2.40 | 2.47 | 2.52 |
| | Random | Random | Random | Random | Random |
| CCr | 0.06 | 0.06 | 0.06 | 0.05 | 0.05 |
| PLr | 2.16 | 2.16 | 2.15 | 2.21 | 2.27 |
| CC / CCr | 5.50 | 5.50 | 5.33 | 6.40 | 6.20 |
| PL / PLr | 1.12 | 1.12 | 1.12 | 1.12 | 1.11 |
| Q | 4.93 | 4.91 | 4.78 | 5.73 | 5.58 |

Source: own calculations. Note: CC stands for clustering coefficient and PL for the average path length. The extension r refers to the respective value calculated from a random network.

The figures in Table 7 are calculated on the basis of the largest network component which encompasses for all observations at least 97% of the nodes which gives the knowledge network a rather cohesive character. Compared to simulated random networks with the same number of ties and nodes, the observed networks have all high clustering coefficients and average path lengths similar to random networks. This combination yields small-world quotients $(Q)$ that are much greater than 1.0 which signifies that the networks can indeed be labeled as small-world networks. A small-world network can be developed from a regular network by adding a number of shortcuts (cf. Figure 6). For a knowledge network based on the co-occurrence of IPC-classes, this means that extra cluster ties are formed, i.e. ties between densely interconnected technology fields. Such new combinations between technology fields indicate a high potential for innovations. In line with this interpretation of the small world-characteristic, an increase in the small-world quotient $(Q)$ signifies an increase in potential innovativeness of actors owing this technological knowledge-base.

### 9.3.3 Tie Strength and Network Overlaps

In social networks, ties can be strong, for instance when we collaborate with someone on a regular basis, or weak, if we meet someone only occasionally. This idea was taken up by Granovetter (1973) who investigates strong and weak ties in a professional context, suggesting that it is weak social ties which are most valuable in the search for employment. In the context of innovation networks, participants benefit from their relational and structural embeddedness, i.e. from direct and indirect linkages to other network participants (chapter 2.1). While strong (direct) ties allow for the exchange of complex information and tacit knowledge due to the possibilities of further inquiries, weak ties, instead, enable the network actors to access entirely new knowledge. Weak ties connect actors to remote subgroups in the innovation network where – with a higher probability – new knowledge can be grasped (Granovetter, 1973; Granovetter, 1983; Rowley, Behrens and Krackhardt, 2000). From a slightly different angle we can also say that a strong tie network is conducive to the diffusion of existing knowledge. In addition, the transfer of tacit knowledge is accelerated in strong tie networks since the strong redundant ties are an indicator for the high level of trustworthiness in the network. On the other hand, weak tie networks are more beneficial for explorative tasks, i.e. the generation of new knowledge which is limited in dense networks in which redundant knowledge supersedes (Rowley, Behrens and Krackhardt, 2000).

Granovetter describes the strength of a tie by the following definition: "The strength of a tie is a (probably linear) combination of the amount of time, the emotional intensity, the intimacy (mutual confiding), and the reciprocal services which characterize the tie" (Granovetter, 1973, p. 1361). Most people would probably intuitively agree on these characteristics to be applied in social interpersonal networks. However, the nodes I analyze in this chapter are knowledge-elements (IPC sub-classes) which I assume to be linked if they co-occur on the same patent. Thus, the strength of a tie must be evaluated by a different characteristic compared to an interpersonal network. To determine the strength of a tie in a knowledge network which illustrates the relatedness of knowledge, I suggest counting the number of co-occurrences of IPC sub-classes (4-digit level) on all the patents in a firm or industry knowledge-base. I consider this an analogy to Granovetter's "time commitment" reasoning. In this vein, I not only analyze the network of a single firm, but I scrutinize the characteristics of the sampled industry knowledge-base which consists of all patents of the 153 firms purposely selected as the sample of analysis.

A central hypothesis of Granovetter's (1973) theory is that the ego-networks of any two actors $i$ and $j$ have a strong tendency for overlapping if the tie between node $i$ and $j$ is a strong tie. That is, if two actors are connected by a strong tie they supposedly share other cooperation partners. Thereby, the dyadic tie structure gets linked into larger network configurations. According to Granovetter (1973), there are two mechanisms which are causal for the appearance of such overlapping structures: First, if we assume that node $i$ is connected with $j$ and with $k$, then the probability for the establishment of a tie between $j$ and $k$ is a function of the tie strength between $i$ and $j$. The higher the frequency of interaction and thus the tie strength between $i$ and $j$ is, the more often also $j$ and $k$ have the opportunity to meet which is conducive to the formation of a tie between $j$ and $k$ (Homan, 1951). The opportunity to meet is particularly elevated once there is not only a strong tie between $i$ and $j$ but also between $i$ and $k$. Second, the more similar two (or more) individuals or organizations are in one or more characteristics, the stronger is the tie between them (homophily theory). Accordingly, if $i$ and $j$ and respectively $i$ and $k$ are connected by strong ties, then $j$ and $k$ are probably also similar, paving the way for the two actors to form a tie (Figure 19).

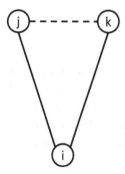

**Figure 19:** Overlapping ego networks (Source: own illustration).

The suggestion that the degree of ego network overlap can be explained by the tie strength of social relations gets in a next step transferred to the patent network representing the knowledge-base of the analyzed automotive firms. I suggest that a tie between two IPC sub-classes (4-digit level) exists whenever two sub-classes co-occur on the same patent. The frequency of co-occurrence of patent classes on a patent reflects the strength of relationship between these knowledge fields as well as their distance (e.g. Schoen et al., 2012). For the mechanisms which are causal for the emergence of overlaps I make the following suggestions: First, if the patent sub-class $i$ is similar, related or complementary and thus regularly co-occurs together with the sub-class $j$ (and $k$), then there is a high probability that the knowledge reflected by the IPC sub-classes $j$ and $k$ is also similar, related or complementary and will hence be combined. Second, the more often the patent sub-classes $i$ and $j$ as well as $i$ and $k$ co-occur on a patent, the higher is the probability that also $j$ and $k$ co-occur on a third patent or that the three together will occur on a fourth patent (Figure 20). The reason is that classes which are technologically-wise similar will often be called for in tandem by EPO examiners.

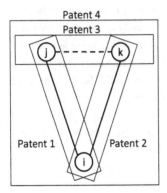

**Figure 20:** Overlapping technology networks (Source: own illustration).

Based on these considerations, I state the following hypothesis: The stronger the tie between two IPC sub-classes is, the higher is the probability that the ego networks of the two sub-classes overlap. The tie strength ($s$) is measured by the number two IPC sub-classes ($i,j$) (4-digit level) co-occur on a patent ($p$) in the sample. The result is a node-by-node matrix which includes the strength of ties among all pairs of actors (named STRENGTH):

$$s_{ij} = \sum p_{ij} \qquad (13)$$

The overlap is operationalized by counting the number of IPC sub-classes both focal nodes ($i,j$) are linked to. These numbers are also arranged as a node-by-node matrix named OVERLAP. For the computation of the overlap the (symmetric) adjacency matrix of ties between IPC sub-classes is dichotomized in a first step which means that the information about the tie strength is discarded. This procedure results in a matrix called $A$. Subsequently, the matrix $A$ gets multiplied by its transpose $A'$ which gives the OVERLAP matrix. Thereby, the number of times each pair of rows in matrix $A$ has a "1" in the same column is counted. A "1" in the same column means that the two IPC sub-classes have a tie to the same third party IPC sub-class. Thus:

$$OVERLAP = A \cdot A' \qquad (14)$$

In this way, I construct a model which is used for the hypothesis test. The variables in the model are matrices including information about the tie existence, strength and the degree of overlap. Thus, I test if the overlap (dependent variable) is correlated with the tie strength (independent variable) in a way that the overlap can be explained by tie strength (for this approach see Borgatti and Feld, 1994). A straightforward way to test this interdependency could possibly be the application of OLS (Ordinary Least Square) regression or rather logistic regression for the case of binary data. However, as in the case of modeling innovation network evolution problems related to statistical methods arise:

- First, we do not have random samples but we are dealing with an entire "population" which is the knowledge network as such.
- Second, variables are probably not drawn from a normal distribution (see the skewed degree distributions). Without an assumption of a distribution of the "population" we do not know against what we can compare the test statistic.
- Third, we are faced with the problem frequently occurring in networks, namely that observations (ties) are not independent, for instance due to the propensity to form closed triads causing trouble with autocorrelation. That is, observations within rows or columns in the matrices tend to be strongly correlated. Strong correlations across observations lead to errors which are correlated with each other. Consequently, standard errors get wrong. Typically correlations result in too small standard errors and thus too small p-values which imply a rejection of the null hypothesis stating that there is no positive relationship between the tie strength and the overlap.

A possible way out is to explicitly model dependency which is done with the stochastic actor-based model for network dynamics (chapter 9.5). Another possible – and for the presented context more obvious option – is to work with a randomization or permutation method of a correlation test which can cope with the reported problems. This approach works as follows: The Pearson correlation between the matrices STRENGTH and OVERLAP is calculated. In order to check the significance of the correlation the following question is raised: How likely is it that we can get a correlation as high as the observed one just by chance? To answer this question the observation is compared against a distribution of correlations from which we know that the process which assigns values to the independent variable matrix is a random process (sampling). Based on this randomization procedure, which is actually a permutation process repeatedly shuffling the values of the original matrix, for each permutation a correlation is calculated resulting in a distribution of random correlation values. These randomized datasets constitute a sampling distribution which resembles the original dataset (automorphism to the original data on a variable). Now we can calculate a p-value which gives the proportion of permutated (random) correlations which are as large as the observed correlations:

$$p = \frac{(Nbr\ permu\ cor \geq obs\ cor) + 1}{Nbr\ permutations + 1} \tag{15}$$

With the permutated matrix a random estimation of the depended variable can be done. The share of coefficients is accounted which is as extreme as the coefficient computed from the observation. If it is located at an extreme high or low percentile (dependent on the chosen significance level) the null hypothesis will be rejected.

The permutation as such is done with the so-called *Quadratic Assignment Procedure (QAP)* which is implemented in the Ucinet software (Borgatti, Everett and Freeman, 2002) (for an overview see Hubert, 1987). QAP is a non-parametric method which mixes (only) the dependent variable data by a sequence of permutations. Dependence within rows and columns is conserved but the relationship between dependent and independent variables is resolved by random permutations of the rows and columns. Values sharing a row/column in the original data set share a row/column in the permuted data as QAP applies the same permutation for the rows as for the columns. After the permutation, we can expect that there is no relation whatsoever between the dependent and the independent variable anymore, which corresponds with the null hypothesis. One advantage of this method is that it does not require an assumption about the distribution of parameters. For the number of permutations a value as large as 5000 is chosen. In general, a high number of permutations improves the estimates of standard error and significance. Krackhardt (1987, 1998) shows that for the case of structural data the QAP method delivers better results for significance tests compared to OLS regressions.

**Table 8:** Pearson correlation between tie strength and network overlap.

|                       | 1998-2002 | 1999-2003 | 2000-2004 | 2001-2005 | 2002-2006 |
|-----------------------|-----------|-----------|-----------|-----------|-----------|
| Pearson correlation   | 0.371     | 0.371     | 0.380     | 0.386     | 0.393     |
| p-value               | 0.0002    | 0.0002    | 0.0002    | 0.0002    | 0.0002    |

Source: own calculations.

For all network observations the Pearson correlation coefficient is greater than 0.37 indicating that there is indeed a positive relationship between the tie strength and an overlap of ego networks (Table 8). Even though this is not a test for the direction of correlation or for causality, the theoretical considerations made before support the reasoning that the stronger the tie is the more the ego networks are overlapping. Moreover, the measured correlation is highly significant as p-values, i.e. the proportion of correlation generated by permutation that are as large as the observed correlation, is very low (< 0.02%).

To improve the validity of this result, Borgatti and Feld (1994) suggest conducting the same test with the non-overlap matrix, i.e. to control the non-overlap matrix with the strength matrix for correlation. Given a positive correlation between the overlap and the strength matrix, a negative correlation between the non-overlap and the strength matrix would be the strongest confirmation for the result calculated before. Contrariwise, a strong positive correlation between the non-overlap and the strength matrix would imply that strong ties most likely occur between nodes that have generally a high number of ties, i.e. their neighborhoods both overlap and do not overlap much. For the presented dataset correlations between the non-overlap and the strength matrix are positive but very small (around 0.04). This result takes me to the conclusion that there is indeed a tendency for nodes which have strong ties to have overlapping ego-networks.

## 9.4     Descriptive Network Statistics

In this subchapter, I report basic statistics which provide a first impression of the observed network evolution of the analyzed automotive innovation network. This impression is supported by a visual mapping of the network change. In addition, I control if the analyzed network exhibits small-world properties, a characteristic which is often found in "natural" networks that are not designed by virtue of a certain policy.

### 9.4.1   Tie Evolution over Time

Figure 21 and Table 9 indicate an increase in the number of established ties between the observation years 2005 and 2006 as well as between the years 2006 and 2007. This can to some extent be explained by an increased number of subsidized research projects as this policy instrument gained in importance over the years (Figure 22). The number of disrupted as well as the number of stable ties is faltering over the observation period. Figure 23 shows the resulting networks for the six observation points (December) 2002-2007 and their regional clustering.

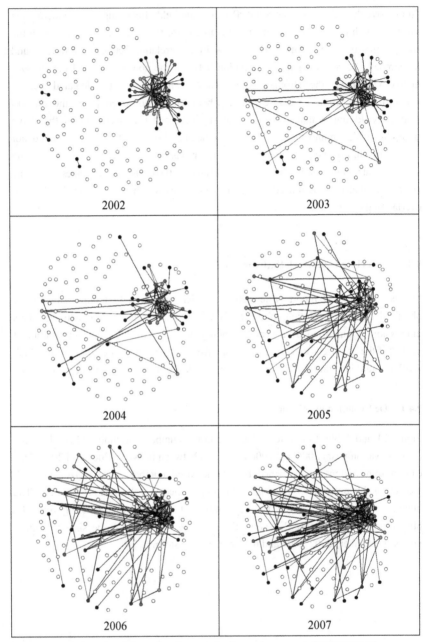

**Figure 21:** Development of the analyzed automotive innovation network (2002-2007) (Source: own illustration).

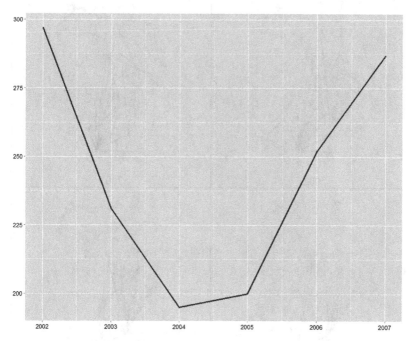

**Figure 22:** Number of projects (Source: own calculation and illustration).

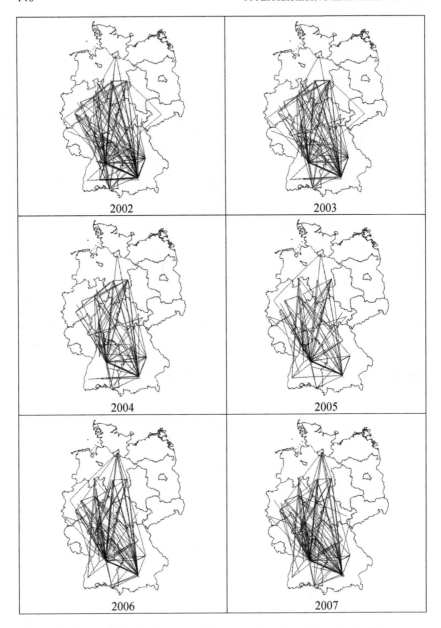

**Figure 23:** Geographical development of the automotive innovation network (Source: own illustration).

**Table 9:** Link development 2002-2007.

| Observation | $0 \rightarrow 0$ | $0 \rightarrow 1$ | $1 \rightarrow 0$ | $1 \rightarrow 1$ |
|---|---|---|---|---|
| 2002 $\rightarrow$ 2003 | 11448 | 14 | 43 | 123 |
| 2003 $\rightarrow$ 2004 | 11481 | 10 | 39 | 98 |
| 2004 $\rightarrow$ 2005 | 11475 | 45 | 58 | 50 |
| 2005 $\rightarrow$ 2006 | 11456 | 77 | 17 | 78 |
| 2006 $\rightarrow$ 2007 | 11440 | 33 | 11 | 144 |

Source: own calculations. Note: The table shows the tie development between observation points. $0 \rightarrow 1$ refers to established, ties, $1 \rightarrow 0$ to disrupted ties and $1 \rightarrow 1$ gives the number of stable ties.

Table 10 shows that the density of the network is overall relatively low. It is slightly diminishing from 2002 to 2005 and then rising again to the final year 2007. Likewise, the average degree centrality which indicates the average number of established cooperative relations is decreasing in the first half and increasing again in the second half. This tendency is confirmed by the number of ties which have been formed in the network (Table 9).

**Table 10:** Density measure 2002-2007.

| Observation | 2002 | 2003 | 2004 | 2005 | 2006 | 2007 |
|---|---|---|---|---|---|---|
| Density | 0.014 | 0.012 | 0.009 | 0.008 | 0.013 | 0.015 |
| Average Degree Centrality | 2.170 | 1.791 | 1.412 | 1.242 | 2.026 | 2.314 |
| Number of ties | 166 | 137 | 108 | 95 | 155 | 177 |

Source: own calculations.

### 9.4.2   Test for Small-World Properties

A potential shortcoming of this network data is a possible political determination, i.e. networks are to some extent designed by political decisions to support certain key technologies that are considered as relevant for the improvement of the competitiveness of the national economy. This includes also granting schemes that disqualify certain firms and reduce the number of eligible firms. Innovation networks generated by policy instruments might differ from emerging networks without external stimulus and confine the interpretation of the results (Schön and Pyka, 2012). Because publicly funded networks (formally) dissolve per definition after the funding period, long lasting linkages for knowledge transfer and learning might not appear.

In many cases, self-organizing networks are characterized by small-world properties (Watts and Strogatz, 1998) which do not appear as frequently in networks created by policy instruments. For instance, small-world properties are found by Uzzi and Spiro (2005) for a network of Broadway musical artists, by Newman (2001b) for networks of scientific co-authoring in seven different scientific disciplines, by Fleming, King and Juda (2007) for patent collaboration networks, by Davis, Yoo and Baker (2003) for the network of US company directors and by Pyka, Gilbert and Ahrweiler (2007) for innovation networks in the biopharmaceutical industries.

The identification of small-world attributes in the observed innovation networks in the automotive industry would weaken the objection towards publicly funded networks. The measurement of the required path length ($PL$) makes only sense in networks where all actors have at least one tie. Therefore, the largest component is extracted from the full network. If the small-world quotient is greater than 1.0, then the network can be designated as small-world network (cf. chapter 9.3.2). The reported statistics by Kogut and Walker (2001) indicate that the critical value of the small-world quotient $Q$ is supposed to increase with rising numbers of nodes in the network. Table 11 shows that $Q$ is in fact for all observed networks larger than 1.0.

**Table 11:** Small-world test.

|  | Network 2002 | Network 2003 | Network 2004 | Network 2005 | Network 2006 | Network 2007 |
|---|---|---|---|---|---|---|
| Nodes (largest component) | 56 | 52 | 46 | 48 | 55 | 64 |
| Ties | 166 | 137 | 108 | 95 | 155 | 177 |
| CC | 0.38 | 0.42 | 0.44 | 0.32 | 0.45 | 0.44 |
| PL | 2.42 | 2.55 | 2.70 | 2.85 | 2.73 | 2.70 |
|  | Random | Random | Random | Random | Random | Random |
| CCr | 0.11 | 0.11 | 0.10 | 0.08 | 0.10 | 0.09 |
| PLr | 2.42 | 2.52 | 2.59 | 2.77 | 2.47 | 2.58 |
| CC / CCr | 3.45 | 3.82 | 4.40 | 4.00 | 4.50 | 4.89 |
| PL / PLr | 1.00 | 1.01 | 1.04 | 1.03 | 1.11 | 1.05 |
| Q | 3.45 | 3.77 | 4.22 | 3.89 | 4.07 | 4.67 |

Source: own calculations. Note: CC stands for clustering coefficient and PL for the average path length. The extension r refers to the respective value calculated from a random network.

The concept of the small-world network (Watts and Strogatz, 1998) was originally developed for one-mode networks such as friendship networks. The interpretation of the clustering coefficient ($CC$), which is required to calculate the small-world quotient

(Q), entails a possible pitfall in the case of a two-mode (affiliation) network which is rarely discussed. For instance, the ties of a two-mode network may be based on the participation and presumed interaction within a common (research-) project. Accordingly, there are two levels of possible nodes for the analysis, namely the level of projects and the level of actors within the projects. In order to analyze a two-mode network it gets usually transformed by projection into a one-mode network as most network measures are only defined for the one-mode case. By projecting a two-mode network into a one-mode network the information of the two-mode structure is lost.[12] As shown in Figure 24, a feature of projected networks is that all project members of the example form a fully linked clique. Since the global network is formed out of the cliques that are linked to each other by firms which participate in multiple projects, the projected one-mode network overstates the true level of clustering compared to a respective random network (Uzzi and Spiro, 2005). Thus, there is a possible pitfall in the interpretation of network measures such as the clustering coefficient.

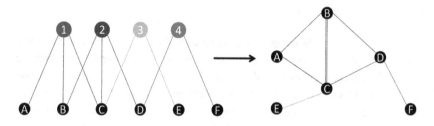

**Figure 24:** Projection of a two-mode-network (Source: own illustration).

The clustering coefficient (CC) is defined as the probability that a connected triplet of nodes is actually a triangle. That is, with the clustering coefficient we can measure the likelihood that two firms which have a common cooperation partner also cooperate among each other. Accordingly, the calculation of a clustering coefficient does not make sense for the case of a two-mode network as it is based on the enumeration of closed triangles in the network:

$$CC = \frac{3 \cdot no.\,of\ triangles}{no.\,of\ triplets} \qquad (16)$$

---

[12] The information of the two-mode structure can however be used to endow the ties with weights. This can be based on the account of common nodes or rather a discounting procedure can be used to devaluate ties. This could for instance mean that ties in projects that include lots of firms get a lower weight than ties in projects that include only few firms. The logic behind this is that in larger groups interaction will be less frequent than in smaller groups (see Newman (2001a) for the discounting approach).

The projected network contains more triangles compared to a typical network which is shaped by a similar tendency for triadic closure. Moreover, the projections may also lead to an artificially high density. Accordingly, the measured clustering coefficient can be a misleading indicator if it is not interpreted with care. It requires some kind of adjustment – at least mentally. To adjust the measure, the suggestion is made that the within-project clustering needs to be deducted. This is related to the observation that a value of the $CC$-ratio (measured $CC$ / random $CC$) in the two-mode case of about 1.0 signifies that the measured clustering in the network comes predominantly from within project clustering and only to a small extent from the ties linking projects. Once the $CC$-ratio grows larger than 1.0 the amount of inter-project clustering starts to grow. A further observation is that with rising $CC$- ratios it is the firms that previously already participated in the same project and had common third party ties which build the links between clusters (Newman, 2001b). In other words, the more a two-mode network gains the character of a small-world network, the more links between clusters are established which allows for a widespread distribution of knowledge nested in research projects throughout the network.

In order to evaluate a clustering coefficient calculated for a one-mode projection of a two-mode innovation network, it is helpful to recapitulate the context in which the measure is applied. It serves as a subindicator for the small-worldness of an innovation network and the small-world character contributes to the knowledge diffusion process throughout the network (Cowan and Jonard, 2003; Cowan and Jonard, 2004). For the analyzed network, the $CC$-ratio is for all observation points considerably larger than 1.0, ranging from 3.45 to 4.67. Furthermore, Table 12 indicates the percentage of firms for the six observation points which participated in more than one project (among those that participated at least in one project) thereby establishing inter-clique ties leading to "real" clustering. The share ranges from 43% to 53% which designates about half of the firms as inter-tie spanners.

**Table 12:** Share of firms in multiple projects.

| 2002 | 2003 | 2004 | 2005 | 2006 | 2007 |
|------|------|------|------|------|------|
| 49 % | 48 % | 43 % | 43 % | 53 % | 52 % |

Source: own calculations.

In addition, the formation of triadic structures in actor-based networks can be related to a social cohesion effect (see the chapter on transitivity) and is a hypothesized property of the observed network. Müller, Buchmann and Kudic (2013) develop a simulation algorithm for network evolution which explicitly takes the tendency for transitive closure into account. Based on this algorithm, networks are simulated with the same number of nodes and ties as in the case of the observed networks. The clustering coefficient of the simulated networks is significantly higher compared to random networks but somewhat lower compared to the observed networks. This brings me to the conclusion that the high clustering coefficient can partly be explained by the projection but partly also by real transitivity based on social interaction patterns. The analysis of two-mode networks became increasingly popular during the last years. New attempts have been made for capturing the degree of clustering more properly in such networks. Basically two different lines are followed: First, an adjustment of the measured clustering coefficient based on the distribution of clique spanning ties (Newman, Strogatz and Watts, 2001; Uzzi and Spiro, 2005). Second, the development of an alternative clustering coefficient measure which can be applied to two-mode networks to avoid projection.

## 9.5 A Stochastic Actor-Based Model for Network Evolution

To model network evolution and to test hypotheses referring to its drivers, I apply a *stochastic actor-based model for network dynamics* (Snijders, 1996; Snijders, 2001; Snijders, 2005) suited for statistical inference analysis based on longitudinal network data. While this model belongs to the class of *agent-based models* (chapter 6.2), the notion *actor-based* or *stochastic actor-oriented model* (SAOM) is used to avoid any misleading association, for instance with principal-agent theory. Actor orientation means that for each change in the network structure the perspective of the focal actor is taken whose tie is changing. The applied model has the advantage of capturing network evolution driven by a combination of effects (independent variables) simultaneously. The model allows for testing hypotheses about driving factors and for estimating parameters. Accordingly, stochastic actor-based models for network dynamics allow analyzing the process of innovation network evolution and disentangling in this complex process different independent variables. Contrariwise, standard regression models can hardly be applied to network data since the independence of observations (tie formation and dissolution) is a prerequisite, whereas tie dependency (endogeneity) is explicitly modeled in the applied approach. A further way to deal with endogeneity is to apply a permutation test, e.g. a QAP (Quadratic Assignment Procedure) test which calculates Pearson correlations between two matrices (cf. chapter 9.3.3). The SAOM could originally only be applied to directed networks. Only recently it has been extended in a way that we can use it for the analysis of undirected (innovation) networks, too (Snijders, 2008). The original intention was to model the evolution of networks in groups consisting of individuals such as school classes, and a main interest was to discover sociological principles of network evolution. Economists and management focused researchers found that stochastic actor-based models are also suitable to capture the evolution of interorganizational and in particular interfirm networks.

Other network models which could also qualify for an application to innovation networks, such as Bala and Goyal (2000) or Marsili, Vega-Redondo and Slanina (2004), are likewise of the agent-based type but they only take a single social theory into account. The widespread used application of scale-free networks (Barabasi and Albert, 1999) is limited because it considers only one explanatory variable, namely the uneven distribution of the actors' degrees. The models of Watts and Strogatz (1998) or Barabasi and Albert (1999) are similarly restricted in their number of considered driving forces and do not model the dissolution of ties. To understand network dynamics, the exclusive focus on the emergence of the network ties is not sufficient. Of course, also the dissolution of network linkages shapes network evolution and provides information about preferences with regard to cooperation partner selection. Here we find the decisive advantage of the stochastic actor-based approach. This model explicitly considers formation and dissolution of network ties and allows for consulting a broad set of explanatory variables.

### 9.5.1  Model Building Blocks

Based on the first observation, the SAOM simulates with a stochastic process networks in the space of all possible networks with the aim to arrive at the structure of the second observation. There are in general a large number of possible networks which have the same number of nodes and ties. What we do not know is which processes lead to the formation of the observed structure. Consequently, the goal is to find a model that represents as accurate as possible the processes that shaped the observed network and to find parameters which make model statistics fit to the observed network statistics. If the modeled network and the observed network share many characteristics, we can conclude that the effects (independent variables) the model contains determined the evolution of the observed network. Also, the statistical significance of estimated parameters is controlled to see if a network configuration resulting from an effect is in the observed network included to an extent that is greater or smaller than what is expected to occur from a random process. The influence of a particular effect is defined by the value of its parameter. In cases where the parameter is large and positive, the corresponding configuration (e.g. the number of closed triads) appears more frequently than for low parameter values. For instance, high values of the parameter that is related to triadic closure let us expect to observe a lot of closed triads. However, due to different scaling, comparisons in the magnitude between effects and especially between models are difficult. To have a single model definition, the number of effect parameters gets reduced to one parameter for each effect, e.g. the assumption is made that the parameters for triadic closure are the same for all closed triads in the network.

In the SAOM, firms are modeled as organizations following decision rules, seeking to improve the value of a so-called *objective function* which reflects their preferences for particular partners and network structures. When an actor $i$ gets the chance to change a tie, it will select the change which yields the highest increase in the objective function (plus a random term). That is, the objective function reflects the value an actor attaches to a certain network structure. The behavior of firms can be described by what Nelson and Winter (1982) call *routines*. In this understanding, firm activities are steered by rather persistent rules that express what a firm does in fields such as production, logistics, R&D etc. "Routines play the role that genes play in biological evolutionary theory" (Nelson and Winter, 1982, p. 14). They are heritable and selectable like genes. Routines are persistent but not static. The direction of change is led by a change routine that reflects a research processes and subsequent change of routines in an evolutionary economic model is comparable to mutation in evolutionary biology. The search process together with the selection mechanism, both are interaction factors changing firm characteristics in an adaptive manner. Such change processes take time. Accordingly, time is naturally an ingredient in a model that reflects adaptive routinized behavior.

The basic network denotation applied with the SAOM is in line with regular social network analysis (e.g. Wasserman and Faust, 1994). A network is represented by an $n$ * $n$ adjacency matrix $X(t_m) = \left( X_{ij}(t_m) \right)$ for m = 1,.., M with $i$ and $j$ ranging from 1 to $g$ which is the number of network actors (nodes). $X_{ij}(t)$ takes the value 0 if there is no link at point $t$ from $i$ to $j$ or 1 if there is a tie from $i$ to $j$ at time $t$. The diagonal of the matrix takes the value 0, $X_{ii}(t) = 0$ for all $i$ as it does not make any sense to a have a tie from an actor to itself. Changes in tie variables are the dependent variables in the model. As an example, Figure 25 shows the formation of a tie between actor $B$ and $C$, and the respective change in the adjacency matrix.

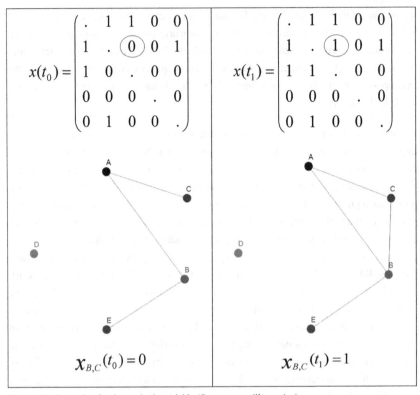

$$x(t_0) = \begin{pmatrix} . & 1 & 1 & 0 & 0 \\ 1 & . & ⓪ & 0 & 1 \\ 1 & 0 & . & 0 & 0 \\ 0 & 0 & 0 & . & 0 \\ 0 & 1 & 0 & 0 & . \end{pmatrix} \qquad x(t_1) = \begin{pmatrix} . & 1 & 1 & 0 & 0 \\ 1 & . & ① & 0 & 1 \\ 1 & 1 & . & 0 & 0 \\ 0 & 0 & 0 & . & 0 \\ 0 & 1 & 0 & 0 & . \end{pmatrix}$$

$$x_{B,C}(t_0) = 0 \qquad\qquad x_{B,C}(t_1) = 1$$

**Figure 25:** Example of a change in tie variable (Source: own illustration).

The SAOM is characterized by two important features: First, it is dynamic in the sense that the ties between the actors change over time and each firm can decide with whom to start or to cease collaboration (control of outgoing tie). This decision is made on the basis of actor characteristics (e.g. homophily) and on the existing respectively possibly realized network structure. The idea that not only individual covariates but also structural network characteristics play a role can be traced back to the concept of *structural individualism* (Udehn, 2002; Hedström, 2005). The emergence of structural effects signifies that ties between two actors (*i,j*) are highly depended on the existence of ties between other actors. Second, network ties are regarded as states, i.e. ties are rather persistent in time and do not represent short events only. Cooperation for innovation has necessarily a long term perspective to match knowledge-bases and tacit knowledge. Partnerships which are inclined for the long run are more stable as they allow for a matching of objectives (Dodgson, 1993). Projects listed in the "Förderkatalog" typically run for at least three years (often longer).

As an approximation, change in the network structure is interpreted as the result of a *continuous-time Markov chain* in the space of all possible networks, meaning that the momentary state of the network determines probabilistically its further evolution. Such models are adequately realistic and qualify for an implementation as a computer simulation for parameter estimation (Snijders, 2005). Simulating network evolution with empirical data requires at least two observations of the network ($M \geq 2$) from which change is taking place between the observation points. A model extension to weighted ties is part of current research only such that at the current state of model development dichotomized adjacency matrices are used.

### 9.5.2  Networks as States of a Markov Chain

The aim is to capture and explain tie changes with a statistical model. *Markov Chain Monte Carlo* (MCMC) techniques are suitable to conduct statistical analysis of network evolution implemented as a stochastic simulation model. Parameters are estimated by a MCMC implementation of the methods of moments applying a stochastic approximation algorithm. A set of initial parameters serves as a basis for a simulation of random networks. In a next step, parameter values are adjusted by comparing the simulated networks with the observation. These steps are repeated until the parameter values stabilize and are used for simulating networks that resemble the observation (Snijders, 2001). For the Markov process, $\chi$ is denoted as the arbitrary finite outcome space (all combinatorial possible adjacency matrices with the same number of nodes and ties). Network evolution is assumed to proceed as a stochastic process in the space of all possible networks. It is further assumed that (entire) networks are random variables ($X$) which have a probability distribution that is complex and cannot be approximated by a regular (e.g. normal) distribution. The actually observed network ($x$) is assumed to be drawn from the population of all possible networks that are simulated based on a model that includes the dependent variables which are tested for their significance as drivers of network evolution. The space of possible networks (the population) is huge. For cooperation networks, i.e. undirected networks, the number of possible networks with the same number of nodes can be calculated as follows: Each dyad between firms ($i,j$) can take the value 1 (if $i$ and $j$ cooperate) or 0 (if $i$ and $j$ do not cooperate) in the adjacency matrix resulting in two possible values. The number of possible dyads in a network is determined by the number of actors $n$ and results in $n(n-1)/2$. Dyads with values 0 and 1 can be combined in all possible ways which gives a totality of $2^{\frac{n(n-1)}{2}}$ distinguishable network structures. As an example, for a network which is composed of only $n = 3$ firms, 8 different networks can be thought of (Figure 26). For larger numbers of actors, the number of possible networks increases very rapidly. For the case of the analyzed automotive network with $n = 153$ firms, there are already $2^{11628}$ possible network structures limiting considerably the possibilities for analytic solutions.

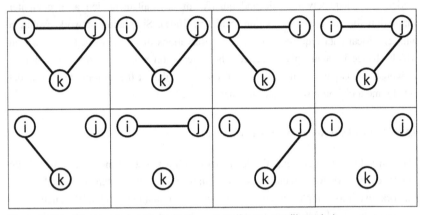

**Figure 26:** Possible network structures with three actors (Source: own illustration).

At least two observations of the network structures need to be made at discrete points in time, but changes are assume to take place unobserved between two observation points and are hence modeled. This kind of analysis is also referred to as network panel analysis (see for the roots of this concept Holland and Leinhardt (1977), Wasserman (1980) and Leenders (1995)). Evolutionary change naturally involves processes taking place in time and the time parameter $t$ is considered a continuous variable. The combination of continuous time in the model and discrete observation points seems to be most adequate for modeling tie emergence and dissolution between actors with ties that endogenously depend upon each other as the following example illustrates: We assume a group of three firms. In the beginning, at $t = 0$, there is no cooperation between them, thus they are fully isolated. At the next observation point, the three firms have started to cooperate among each other and have formed a closed triangle. The emergence of the triangular structure cannot be explained in a model with discrete time only since the triangle might just happen to be there at a certain observation point. The emergence as a step by step process can only be modeled with a continuous time approach.

The number of observation points $t_1$ to $t_M$ is an element in a continuous series of time points $\tau = [t_1, t_M] = \{t \in \mathbb{R} \,|\, t_1 \leq t \leq t_M\}$. The network of the first observation is not simulated but the observation is used as initial structure from which on the change to the second observation is simulated. That is, the simulation seeks to grasp the change which happens unobserved between two consecutive observation points (Figure 27).

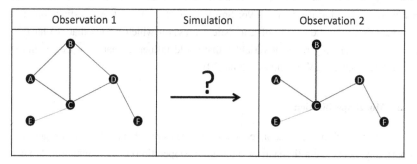

**Figure 27:** Simulation of unobserved change (Source: own illustration).

For a stochastic process $\{X(t)|t \in \tau\}$ to be a Markov process the following condition has to be fulfilled: For any point in time $t_a \in \tau$, the conditional distribution of future states $\{X(t)|t > t_a\}$, given the present and the past $\{X(t)| t \leq t_0\}$, is only a function of the present state $X(t_a)$ (Snijders, 2001). In other words, the current state of the network is the constraint for its further direction and dynamic of change. In this respect, the model is somewhat myopic since the earlier past has no direct effect on the future evolution. We can also speak of myopic stochastic optimization: All the information which drives the change process is incorporated in the current state. In order to make the model more "history friendly" and to remedy to some extent its myopic character, it is advisable to include covariates which represent relevant information of the past (Snijders, 1996). This could for instance be the age or cumulated size of an actor's knowledge-base.

The probability for any possible outcome of a stochastic process $x \in X$ for any pair of time points $t_a < t_b$ is given by:

$$P\{X(t_b) = x|X(t) = x(t) for all t \leq t_a\}$$
$$= P\{X(t_b) = x|X(t_a) = x(t_a)\}$$

(17)

Since the aim is not only to demonstrate and make comprehensible which mechanisms are at work in a network but to estimate parameters for a model, a few data requirements have to be fulfilled. First, we need enough data in order to actually estimate parameters and to ensure a high goodness-of-fit level. Second, parsimony is desired, that is, the model should not include more details (variables) than what can be estimated from the empirical data (Hedström, 2005). Once we have collected enough information about properties and interaction of actors, the model parameters can be estimated. To estimate the parameters, I use the SIENA program which applies the described Markov chain Monte Carlo implementation of the method of moments for parameter estimation (Snijders, 1996; Snijders, 2001).[13] The MCMC estimations can be interpreted like the results of a logistic regression which means that we have to control for potentially relevant variables that could influence change processes in the network (Balland, De Vaan and Boschma, 2012).

### 9.5.3 Model Specification

The first step of the simulation concerns the decision about which actor gets the opportunity to change a tie (start or interrupt a cooperation) or to maintain it. We can either assign the same probability to each actor (same rate of change $\rho$) or make this depended on covariates or network positions. The model which I develop in this dissertation is specified by a uniform rate function. Moreover, I assume that the rate function $\rho_m$ is constant between $t_m$ and $t_{m+1}$. At any point in time, only one actor can change a tie in the model, thus, ties cannot be changed simultaneously by collusion or by other forms of coordination. Actors react to change made in a previous time-step by another network actor (Holland and Leinhardt, 1977; Snijders, 2001). Accordingly, the network evolution process gets subdivided into the smallest possible components and encompasses in fact two subprocesses:

- The first subprocess relates to the regularity an actor gets the chance to change a tie ($\lambda_i(x)$) (rate function).
- The second subprocess refers to changes of the tie status (objective function).

The process of changing a tie variable is called a micro-step. The rate function $\lambda_i(x)$ for actor $i$ determines how regularly the actor does a micro-step. The concept of a micro-step means that two consecutive network states only differ by just one tie variable. A larger change in the network structure from one observation to the next is assumed to accrue from a sequence of micro-steps. The probability for doing a particular step depends in turn on the objective function (18). Tie changes are determined by the objective function which expresses how a firm perceives the network and evaluates the different change options. Regarded from a different angle,

---

[13] The more common maximum likelihood estimation method has also become available. However, reported experience suggests that estimations do not deviate strongly from a method of moments estimation but the estimation process takes considerably longer (Ripley, Snijders and Preciado, 2010).

the objective function represents the preference distribution of an actor over the set $\chi$ of all possible networks. It models the attractiveness of a network state $x$ for actor $i$. In particular, it is used to compare the attractiveness of different tie change options. For any possible state of the network, the objective function takes a certain value. The aim of each actor is to increase the value of the objective function. For higher values the probability increases that the actor opts for the respective network state. For instance, it is counted how many transitive triads are at a certain point in time in the network and how many will be there if the actor $i$ does one of the possible micro steps. Formally, the objective function is a linear combination of a variety of weighted components which are called *effects* (independent variables). The exact configuration depends on the hypotheses which are derived from theory (cf. chapter 7) and its parameters are estimated from the collected data:

$$f_i(x^0, x, v, w) = \sum_k \beta_k s_{ki}(x^0, x, v, w) \tag{18}$$

The value of the objective function $f_i$ for actor $i$ depends on the current state $(x^0)$ of the network, a potential future state $(x)$ as well as on actor attributes $(v)$ and dyadic attributes such as the different distance categories $(w)$. The functions $s_{ki}(x)$ are the effects (independent variables) that are based on previously explained theoretical considerations and represent the operationalized hypotheses. Weights $\beta_k$ are the statistical parameters which have to be estimated. If $\beta_k = 0$, then the corresponding effect does not play a role in network evolution; if $\beta_k > 0$ there is higher probability of moving in the direction where the respective effect is higher. Effects depending on the network are structural or endogenous effects. Effects depending on external attributes are covariates or exogenous effects. Considering the arguments of the objective function, the greater the values of $\beta_k$ are, the higher will be the product of the statistics $s_{ik}(X(t_{m+1}))$ (e.g. number of triads) given a network $x^{obs}(t_m)$. If there is a preference for transitive structures, the firm will choose a step which maximizes the numbers of transitive triads (ceteris paribus the other effects). The choice probability for doing a possible micro-step is proportional to the exponential function of the objective function:

$$\Pr(x \to_i x') \propto \exp(f_i(x')) \tag{19}$$

The tie change routine reflects a myopic stochastic optimization rule. It is myopic because only the state of the network immediately after the micro-step is taken into account. When an actor $i$ changes an outgoing tie $(X_{i1}, \ldots, X_{ig})$ (it is assumed that it has full control over outgoing ties), it will opt for the change which yields the highest increase in the value of the objective function (plus a random term). For a given network state $x = X(t)$ the result of a change of a single tie variable $x_{ij}$ into $1 - x_{ij}$ is denoted as $x(i \sim > j)$. For modeling the change a random variable $U(j)$ needs to be considered which represents the unexplained or residual part of the attraction for $i$ to $j$. $U_j$ is in fact a set of random variables distributed symmetrically around $0$ and independently generated for each new micro-step. As an optimization rule, each actor (who gets the chance to make a change) changes its tie variable with that actor $j$ for whom the value of $f_i\big(x(i \sim > j)\big) + U(j)$ becomes largest (Snijders, 2001).

### 9.5.4 Estimation of Parameters

The model parameters cannot be calculated explicitly but they can be estimated from simulations. The core idea for the estimation is that we have to find values for the parameters $\beta_k$ such that the expected value of the simulated statistics $s_{ik}\big(X(t_{m+1})\big)$ (summarized over $i$ and $m$) equal the observed values of the empirical networks. The test statistics $(s_k)$ are hence effect counts. For instance, it is counted how many closed triads there are in a network. The target value for the simulation can consequently be expressed as:

$$s_k^{obs} \stackrel{!}{=} \sum_{m=1}^{M-1} \sum_{i=1}^{g} s_{ik}\big(x^{obs}(t_{m+1})\big) \ (k = 1, \dots, L) \tag{20}$$

In other words, the aim of the estimation is to fit the expected statistics to the observed statistics by simulating networks which fulfill this condition. This could for example mean that we search for a simulated network which has as many closed triads as the observed network. It is furthermore assumed that $c_m$ is the distance between two networks observed at two consecutive observation points, that is $c_m = |x^{obs}(t_{m+1}) - x^{obs}(t_m)|$. With the simulation algorithm, which is actor-based and uses as its core element the objective function, the aim is to iteratively find a network where $|X_m(R_m) - x^{obs}(t_m)| = c_m$, with $R_m$ as the first point in time where the required condition holds. Based on this network, generated statistics are calculated:

$$S_k = \sum_{m=1}^{M-1} \sum_{i=1}^{g} s_{ik}(X_m(R_m)) \ (k = 1, \dots, L) \tag{21}$$

The global aim is to find a vector $\hat{\beta}$ of parameter values for which the moment equation holds (method of moments), that is, the simulated and the observed vectors of parameters are equal:

$$\varepsilon_{\beta} S = s^{obs} \qquad (22)$$

Due to the complexity of the evolving network, the expected values cannot be calculated directly and are therefore taken from the simulation. The MCMC algorithm is used to approximate the solution of the moment equation (22) in a gradual procedure. To simulate a network which has the envisaged characteristics, a stochastic approximation method is used to approximate moment estimates (Robbins and Monro, 1951). Expected values are approximated as the averages over a lot of simulated networks. Observed values ("target values") are calculated from the empirical data set. For the test of a hypothesis, a test statistic and a distribution of the test statistic according to a null hypothesis are required. T-ratios[14] are approximately normally distributed. Based on this, a p-value for the null hypothesis suggesting that the tested effect has no impact can be calculated ($H_0: \beta_k = 0$). In this way it becomes possible to find out which of the competing effects, formulated as hypotheses, are most probably present in the observed network evolution.

In order to attain proper estimations, it is necessary to have a certain quantity of change between two consecutive observations. On the other side, too much change cannot be handled with the model. As mentioned before, the assumption here is that change in the network takes place in a gradual stepwise procedure rather than by sudden "shocks". In other words, networks are regarded as states rather than short events. The degree of change between consecutive observation points needs to be large enough as tie changes provide the information which is required for estimating parameters of the objective function. Based on experience, there should not be less than 40 changes across all observation points (Snijders, Van de Bunt and Steglich, 2010). More tie changes are welcome as they provide more information about the drivers as long as changes between observations happen only gradually. Thus, it is advisable for studying network evolution to have an idea about how much change is observed. To test for gradual change in the network, the Jaccard index ($J$) (Table 13) is calculated as an indicator ($M_{11}$ = Number of stable links; $M_{01}$ = Number of interrupted links; $M_{10}$ = Number of formed links):

$$J = \frac{M_{11}}{M_{01} + M_{10} + M_{11}} \qquad (23)$$

---

[14] T-ratios are given as $\frac{\beta_k}{sd(\beta_k)}$. With the assumption that t-ratios are approximately normally distributed the significance of effects can be tested (p-values).

**Table 13:** Jaccard index.

| Observation | Jaccard Index |
|---|---|
| 2002 → 2003 | 0.683 |
| 2003 → 2004 | 0.667 |
| 2004 → 2005 | 0.327 |
| 2005 → 2006 | 0.453 |
| 2006 → 2007 | 0.766 |

Source: own calculations.

The value of the Jaccard index should ideally be higher than 0.3 (Snijders, Van de Bunt and Steglich, 2010). This is the case for all observations, making the observations a good basis for the simulation and parameter estimation.

### 9.5.5  Assumptions about Tie Initiation and Dissolution

For each tie there are two actors involved. Thus, there has to be an assumption to cope with the tie formation (and dissolution) process. We can indeed think of three possible tie establishment procedures: First, a tie could be established if either of the two actors wishes to establish it. This possibility is called unilateral imposition (disjunctive). Second, a tie to come into existence could require a mutual agreement (conjunctive). A third way to decide about the emergence of a tie is to calculate a net value, i.e. a positive net value will lead to a tie and the gained value for one actor may compensate a loss for another actor (compensatory). These three options need to be analyzed in tandem with the actor oriented or dyad oriented opportunity element of the rate function. For innovation networks, i.e. the case of undirected ties, three principal tie establishment procedures are plausible (Van de Bunt and Groenewegen, 2007):

- One-sided initiative reciprocal confirmation model (actor oriented): An actor $i$ is given the opportunity to change a tie variable $X_{ij}$. A new tie to an actor $j$ must be proposed and accepted by the potential partner. If the proposal gets denied, the tie will not be established.

- Two-sided opportunity model (dyad oriented and conjunctive): Here, the tie will only be established if it is beneficial for both actors. A pair of actors $i$ and $j$ gets the opportunity to change their tie variable $X_{ij}$, i.e., to start or terminate a cooperation or to keep-up the cooperation.

- Pairwise compensatory (additive and disjunctive) forcing model (dyad oriented): The two actors $i$ and $j$ meet to discuss their tie variable. In this case it is sufficient that the tie is beneficial for one out of the two to become realized. Actors $i$ and $j$ meet and reconsider their tie based on the summed value of the objective function.

I opted for the "one-sided initiative reciprocal confirmation model", which is probably closest to reality, in which one actor takes the initiative and the other one accepts the offer for cooperation or refuses it.

## 9.6    Hypotheses and Operationalization of Concepts

In order to obtain meaningful estimations, it is essential to select effects (independent variables) which reflect the hypothesized mechanisms. As the analyzed networks are undirected, I only operationalize effects which may have a significant impact in the context of undirected networks. There are other important effects which, however, only make sense in directed networks. For instance, reciprocity is a ubiquitous effect which is per definition only meaningful in directed networks. For the case of an undirected network, it is clear that if actor $i$ cooperates with $j$, $j$ also cooperates with $i$. For undirected networks, such as innovation networks, ego and alter effects cannot be differentiated. The ego effect controls if actors with higher values in a specific covariate select more partners. The alter effect measures if actors with higher values in a covariate are more often chosen as a collaboration partner. The similarity effect (homophily effect) is particularly important. It measures if ties are more often established between actors which have similar values in a covariate.

Ideally, we have complete network data to estimate parameters for the model. However, when we analyze firm networks over a period of several years, there will be most probably firms entering the network only after the start of the observations and there will be firms leaving the network due to mergers, bankruptcy etc. before the end of the observations, leading to an unbalanced panel. There are several ways to accommodate for such actor changes in the network. One possibility is to insert so called structural zeros which specify that certain ties cannot emerge since the actors are absent. The most appropriate possibility, which I apply, is to introduce a *composition change matrix*. This matrix contains the information about the points in time when an actor joins or leaves the network (see Table A. 7 for details) (Snijders, 2008).

The core interest is to identify the determinants for the emergence and dissolution of ties in the observed network. From the firm's perspective, I address the question concerning the guidelines that determine the decision to cooperate and to select a cooperation partner. Therefore, I test the relevance of the following factors for network evolution: absorptive capacity, technological distance, knowledge-base modularity, geographical distance, transitivity and experience with cooperation. These factors have been discussed exhaustively in chapter 7 and are at this stage formulated as hypotheses and operationalized for parameter estimation and significance tests.

### 9.6.1 Absorptive Capacity

The formation of alliances is more interesting for firms which have high levels of absorptive capacity as they have better capabilities in evaluating, assimilating and exploiting external knowledge (Cohen and Levinthal, 1990). Empirical studies confirm that firms with high absorptive capacity are more likely to initiate a cooperation (Giuliani and Bell, 2005; Boschma and ter Wal, 2007; Morrison, 2008). For explaining the probability to start or terminate a cooperation, a measure for the absorptive capacity should hence be taken into account. Cohen and Levinthal (1989, 1990) suggest that R&D efforts are an important determinant of a firm's absorptive capacity. That is, the absorptive capacity ($v_{absorpcai(t)}$) is a function of the R&D efforts which I approximate with patent data.

Hypothesis H1: Consequently, I expect firms which have higher levels of absorptive capacity to also have a higher propensity to collaborate as they face more opportunities to benefit from external knowledge.

The absorptive capacity ($v_{absorpca\ i(t)}$) is approximated by taking the natural logarithm of the number of patents ($NbPatents_{i(t:t-5)}$) a firm applied for in the five years prior to the observation point. Accordingly, absorptive capacities of actors increase with diminishing rates with the accumulated patenting activity.

$$v_{absorpca\ i(t)} = \ln{(NbPatents_{i(t:t-5)})} \qquad (24)$$

### 9.6.2 Technological Distance

Hypothesis H2a suggests that firms whose knowledge-bases are more similar (small technological distance) are more inclined to cooperate. In other words, for effective learning, cooperating firms need similar knowledge-bases which reflect a common understanding of problems and increase the capacity to absorb each other's knowledge (Colombo, 2003). On the other hand, invention and innovation can be understood as new combinations of existing knowledge which involves the combination of more dissimilar knowledge-bases. Accordingly, hypothesis H2b states that firms whose knowledge-bases are more dissimilar (large technological distance) are more inclined to cooperate.

For the calculation of distances between firms in the technological knowledge space, I apply the Euclidean distance measure $(E)$ based on a firm's patent portfolio which encompasses all EPO patents filed not more than 5 years prior to the observation point. This approach of calculating a positional vector was developed by Jaffe (1986) and is for instance also applied by Yang, Phelps and Steensma (2010). It assumes that the distribution of a firm's patents across the patent classes of the entire sample reflects the distribution of its technological knowledge. In a first step, a vector is calculated which places each firm in an $N$-dimensional vector space. The number of dimensions $N$ results from the number of 3-digit IPC classes in which all firms filed patents (priority filling). The firm vector $p$ is given by the relative share of patents a firm has in the $N$ patent classes. For instance, if $N$ was only two (e.g. B60 and B29) and a firm has 40% of its patents in class B60 and 60% in B29 the vector would be $(p^{B60}; p^{B29}) = (0.4; 0.6)$. In a second step, differences between vectors representing distances in the technology space are calculated. Thus the technological distance $(w_{techdis\ ij})$ between two firms $i$ and $j$ is calculated as:

$$w_{techdis\ ij} = \sqrt{\sum_{c=1}^{N} (p_i^c - p_j^c)^2} \qquad (25)$$

### 9.6.3 Knowledge-Base Modularity

To analyze the decomposability of a firm's knowledge-base, I consider it as a network of discrete knowledge elements that are linked by patents (affiliation or co-occurrence network). I further assume that a tie emerges between technology classes once they are mentioned on the same patent (Saviotti, 2009). This approach is in line with Yayavaram and Ahuja (2008, p. 334) who state that "the set of couplings or ties together with the strength of the ties constitute the structure of a firm's knowledge base." The connection of knowledge elements is not only dichotomous but varies in its intensity. For instance, two knowledge elements (IPC[15] classes) $A$ and $B$ may appear ten times together on patents of a firm while the elements $A$ and $C$ may only appear once or twice together. Consequently, for the network representing the knowledge-base of a firm, the frequency of co-occurrences is used as weights for the ties.

---

[15] IPC means international patent classification.

As hypothesis 3, I propose that the propensity of two firms to cooperate rises with the degree of modularity of their knowledge-bases. As a measure for the degree of modularity of a knowledge-base, I calculate a slightly modified *clustering coefficient* which I call *clustering indicator* ($ci$). The first step in calculating the modularity of the knowledge-bases is to reconstruct the knowledge-base as a network. I take IPC sub-classes (4-digit level) as nodes and add a tie between nodes whenever the subclasses co-occur on a patent (Saviotti, 2009). Doing this, I reconstruct the knowledge network from patents for each of the analyzed 153 firms in moving time windows each encompassing five years (1998-2002 to 2002-2006). By first neglecting the tie strength (dichotomization of the adjacency matrix), I calculate the clustering coefficient ($cc$) for each node of a firm's knowledge-base network. The clustering coefficient ($cc_i$) for node (IPC sub-class) $i$ with $k_i$ ties is defined in equation 26:

$$cc_i = \frac{n_i}{\frac{k_i \cdot (k_i - 1)}{2}} \qquad (26)$$

The calculation includes $n_i$, the number of ties between the $k_i$ neighbors of node $i$. The denominator represents the maximum number of ties which are possible between the $k_i$ neighbors of node $i$. In order to give weights to the IPC subclasses which appear more often in the patent portfolio, I calculate in equation (27) the share ($rsc_i$) of an IPC sub-class ($c_i$) relative to all IPC sub-classes in the portfolio such as:

$$rsc_i = \frac{C_i}{\sum C_i} \qquad (27)$$

In a final step, the clustering indicator ($ci_i$) for each firm's knowledge-base is calculated in equation (28) by multiplying the clustering coefficients $cc_i$ of each node with the relative shares of the IPC sub-classes ($rsc_i$) and summing them up to weighted clustering coefficients:

$$v_{modular\ i(t)} = ci_j = \sum cc_{ij} \cdot rsc_{ij} \qquad (28)$$

### 9.6.4   Geographical Distance

From the theoretical arguments (chapter 7.5) it follows that co-located firms have a higher propensity to cooperate (hypotheses H4). Figure 28 shows the clustered geographical dispersion of German automotive firms (one dot per firm based on the sample). Most firms are located in the south eastern, south western and western regions of Germany close to Munich, Stuttgart, Frankfurt and in the Ruhr Area. In order to form a pairwise distance matrix, geographical distances between all pairs of actors $(dist_{ij})$ have been retrieved by a specific search routine from the internet navigation service *Google Maps*[16] and logarithmized with the natural logarithm:

$$w^{17}_{geodis\ ij} = \ln(dist_{ij}) \qquad (29)$$

**Figure 28:** Geographical positioning of analyzed automotive firms (Source: own illustration).

---

[16] To calculate distances, I used Google Refine and the Google Maps API with the following expression: "http://maps.googleapis.com/maps/api/directions/json?origin="+escape(cells["Origin"].value, "url")+"&destination="+escape(cells["Destination"].value, "url")+"&sensor=false", and to extract distances from the response: "with(value.parseJson().routes[0].legs[0].distance.pair.pair.text)".

[17] For all dyadic covariates $S_{idyadic} = \sum_j ij(w_{ij} - \bar{w})$.

**9.6.5  Transitivity**

Positive values for the transitivity effect signify that firms which share a common cooperation partner collaborate with a higher probability compared to firms which do not have a partner in common. Groups of strongly interconnected actors generally show a high level of mutual trust which is conducive for cooperation (Walker, Kogut and Shan, 1997; Buskens and Raub, 2002). In this regard, Reagans and McEvily (2003) demonstrate that strong social cohesion around a relationship reinforces the willingness and motivation to invest time, energy and effort in sharing knowledge with others. Trust in dense parts of the network facilitates intensive exchange of complex or sensitive knowledge (Zaheer and Bell, 2005). Accordingly, I state hypothesis H5: Firms sharing a common cooperation partner are more likely to collaborate compared to other actors which do not have a partner in common. Transitivity is measured by the number of closed triplets an actor is involved in[18]:

$$T_i = \sum_{j<k} x_{ij} x_{ik} x_{jk} \qquad (30)$$

**9.6.6  Experience with Cooperation**

I state for hypothesis H6 that a large record of collaborative activities signals a larger attractiveness as well as preference for further collaboration. This reflects that from outside it is rather difficult to scan a firm's valuable resources, in particular its knowledge-base. A firm which has been often involved in cooperative projects signals to be a valuable partner with a good reputation and established routines of collaboration. Cooperation capabilities are specific and not transferable resources which can enhance a firm's ability to identify partners, initiate collaborations and manage the partnerships successfully (e.g. Makadok, 2001). Experienced firms have implemented collaboration management routines to coordinate the portfolio of different types of alliances (Kale, Dyer and Singh, 2002). Developing experience is time consuming because firms are forced to adapt internal routines (Powell, Koput and Smith-Doerr, 1996). However, it is worth the effort as it not only enables a firm to become effectively embedded in a formal innovation network but also paves the ground for likewise important informal collaboration (Pyka, 2000). With equation (31) I measure the experience of a firm ($v_{exp\ i(t)}$) with the frequency of participation in subsidized R&D projects with partners ($NbR\&D_{projects\ i(t:1998)}$) both from within and

---

[18] $X$ takes the value 1 if a tie between two actors is established and 0 if a tie is absent.

outside of the automotive sample starting with the year 1998 which is five years prior to the first observation.

$$v_{exp\ i(t)} = NbR\&D\ projects_{i(t:1998)} \qquad (31)$$

### 9.6.7 Control Variables

I include three controls to the model, referring to a capacity, an experience and a cost effect. Larger firms can coordinate at the same time more cooperation partners than smaller firms. Accordingly, the model needs to control for firm size. For measuring size I distinguish three categories, namely large firms, medium sized firms and small firms. Threshold levels are applied for the number of employees and/or the annual turnover for the years 2002-2010 (e.g. Gulati and Gargiulo, 1999). It also controls for inertia in large firms which makes a knowledge-base more rigid. The required information is taken from the companies' websites and from excerpts of the commercial register (accessed via LexisNexis). Due to some missing data I apply the usual categorization (OECD, 2005):

• Category 1 (Large): > 250 employees; turnover ≥ 50 Mio. €
• Category 2 (Medium): 50-249 employees; turnover < 50 Mio. €
• Category 3 (Small): 10-49 employees; turnover < 10 Mio. €

The second control variable is the industry experience of firms. Older firms are more experienced and can manage a higher number of collaborative projects. To approximate experience I apply the natural logarithm of firm age.

Finally, the density (degree) variable controls for the coordination costs of relations. It indicates why we do not observe fully connected networks in which each firm cooperates with all other firms and needs always to be included (Snijders, Van de Bunt and Steglich, 2010). It represents the balance of benefits and costs of an arbitrary tie. Collaboration is a means to cope with the uncertainty in a technology field. At the same time it creates a new facet of uncertainty which refers to the decision and choice of becoming involved in joint projects with partners. Arbitrary refers to a tie with an actor which is not particularly appealing due to its network position and its characteristics. In reality, most networks have a rather low density which signifies that the costs for connecting to an arbitrary actor are higher than the benefits. In line with these theoretical considerations, very often we find a negative parameter value for this effect.

Density is approximated by the actors' degree centralities:

$$D_i = \sum_i x_{ij}$$                                                                                 ( 32 )

## 9.7    Estimation Results

Model parameters have been estimated with the stochastic actor-based network model as implemented in the SIENA program based on the $R$ platform (Ripley, Snijders and Preciado, 2010). SIENA estimates parameters $\beta_k$ and the respective standard errors for the model. Simulation runs have been repeated 3000 times. A first parameter indicating the goodness-of-fit of the estimated model is the t-value of convergence:

$$t = \frac{E(simulated\ feature) - observed\ feature}{st.dev.(simulated\ feature)}$$                ( 33 )

It indicates the (average) deviation of observed network features (target values) from simulated features based on the estimated model. Convergence is excellent if the t-value is smaller than 0.1, which I found for all variables of the estimated models (Balland, De Vaan and Boschma, 2012). To further improve the model-fit I tested for time heterogeneity in the data and added time dummies to the structural network effects (density and transitive triads) (cf. Figure B. 1). These dummies also capture potential business cycles effects on the network evolution. The rate parameters $\rho$ (Table 14) indicate the estimated number of opportunities for change per actor between two observations. It increases between the observations three and four and falls back to a lower level between the observations five and six. It must be kept in mind that not all opportunities for change will lead to an actual change. A firm might simply prefer to keep the status quo. Accordingly, the average observed number of changes per firms is typically smaller than the estimated rate of unobserved change indicates.

**Table 14:** Rate parameter estimates.

| Observation | Rate parameters Model 1 | Rate parameters Model 2 | Rate parameters Model 3 |
|---|---|---|---|
| 2002 → 2003 | 2.29 | 2.37 | 2.15 |
| 2003 → 2004 | 2.07 | 2.11 | 1.88 |
| 2004 → 2005 | 9.58 | 10.18 | 8.03 |
| 2005 → 2006 | 6.13 | 6.28 | 3.84 |
| 2006 → 2007 | 1.61 | 1.63 | 1.37 |

Source: own calculations.

**Table 15:** Effect parameter estimates.

| Variable | Model 1<br>Value (sd)<br>(knowledge related effects) | Model 2<br>Value (sd)<br>(all effects) | Model 3<br>Value (sd)<br>(all effects & time dummies) |
|---|---|---|---|
| Degree (density) (control) | -2.0680***<br>(0.0483) | -2.0835***<br>(0.0465) | -2.1818***<br>(0.0636) |
| Absorptive capacity | 0.2259***<br>(0.0284) | 0.1325***<br>(0.0353) | 0.1810***<br>(0.0402) |
| Technological distance[19] | -0.5251**<br>(0.2228) | -0.5716***<br>(0.2212) | -0.6586***<br>(0.2433) |
| KB modularity | 0.2397<br>(0.2449) | 0.1500<br>(0.2335) | 0.1210<br>(0.2799) |
| Geographic distance | | -0.1390***<br>(0.0321) | -0.1551***<br>(0.0346) |
| Transitive triads | 0.4049***<br>(0.0281) | 0.4072***<br>(0.0290) | 0.4777***<br>(0.0410) |
| Cooperative experience | | 0.0101***<br>(0.0023) | 0.0026<br>(0.0026) |
| Firm size (control) | | 0.1089<br>(0.0969) | 0.1007<br>(0.1118) |
| Industry experience (control) | | 0.0025<br>(0.0425) | -0.0349<br>(0.0499) |

Source: own calculations. Significance levels: p<0.1*; p<0.05**, p<0.01***.

Table 15 summarizes the resulting coefficient values for the models. Model 1 includes only the knowledge related variables together with the endogenous variable and the density control variable. Model 2 is based on the entire set of derived variables. Model 3 is based on Model 2 complemented by time dummies. For the hypotheses tests, the null hypothesis H0 states that the respective covariate does not affect network change. In the simplest model (model 1), the effects are significant except the knowledge-base modularity effect. For the more complex models, significance does not change a lot. The additional covariates are significant except the capacity effect measured by firm size and industry experience. In model 3, the cooperative experience effect is either not significant.

The estimated objective function for model 3 is:

$$
\begin{aligned}
f_i = &-2.1818s_{i1}(x) + 0.1810s_{i2}(x) - 0.6586s_{i3}(x) + 0.1210s_{i4}(x) \\
&- 0.1551s_{i5}(x) + 0.4777s_{i6}(x) + 0.0026s_{i7}(x) \\
&+ 0.1007s_{i8}(x) - 0.0349s_{i10}(x)
\end{aligned}
\tag{34}
$$

---

[19] I also added a squared distance term to control for a curvilinear relationship. This, however, leads to insignificant results.

Hypothesis H1 deals with the positive effects of the absorptive capacity with respect to the propensity to collaborate. The estimations confirm that firms with broad absorptive capacities are likely to benefit more from cooperation and therefore more intensely engage in networks. Firms which have a larger knowledge-base have more incentives to cooperate as they are better capable of making use of the other firm's knowledge-base they get access to.

For the technological distance I find a negative parameter value which suggests that there is a tendency for firms with similar knowledge-bases to cooperate (Hypothesis H2a). Furthermore, there are various ways of operationalizing the concept of technological distance (cf. Benner and Waldfogel, 2008) and this factor provides room for further investigation.

The parameter for the knowledge-base modularity is positive but not significant. So, there is only weak indication that the structure of the knowledge-base is a determinant for the attractiveness of becoming a collaboration partner and for engaging in cooperation. The results do not confirm hypothesis H3 about the beneficial effects of a modular knowledge structure which facilitates recombinatorial research and with it possibilities to benefit from sharing knowledge in innovation networks. More research is required on this point.

Hypothesis H4 indicates an inverse relationship between the propensity to cooperate and the geographical distance. The parameter for geographic distance is negative and highly significant. This indicates that ties emerge more frequently between firms that are located in relative geographical proximity compared to more distant firms. From this follows that geographical distance is an important factor in the automotive industry.

Hypothesis H5 which suggests a high cliquishness among network partners is confirmed. This indicates a significant endogenous network effect leading to the formation of cohesive triadic subgroups caused by trusted partnerships.

A further tested covariate is the experience in cooperation. The results confirm hypotheses H6 for model 2 suggesting that firms with more experience in cooperation are more open to participate in collaboration projects. However, the effect becomes insignificant once time dummies are added.

Additionally, the results indicate that firm age has no significant influence on the handling of collaborative projects. The impact of firm size on collaboration is not visible either. There is no clear effect supporting either small or large firms with respect to their collaboration activities. This is in fact a positive result for innovation policy makers since small and young firms do not seem to be restrained form participation in collaborative research projects. The effect for density is negative and highly significant which indicates that there are cooperation costs which inhibit firms to start too many collaborative projects.

The correlations between estimates are used to check for collinearity between effects (cf. Table A. 3, Table A. and Table A. 5 ). Collinearity refers to possibly different combinations of parameter values representing the same data pattern, that is, the same values of the network statistics such as the number of triads. Near collinearity is reached if the correlations are very close to +1.0 or -1.0. This does however not mean that some of the tested effects should rather be neglected but that there is a trade-off between highly correlated effects. Eventually the selection of a model should be based on theoretical considerations, the questions at stake and experience. While this is often straightforward for covariate or dyadic effects, structural effects are less intuitive to be implemented but likewise important for a satisfying model-fit. The correlations between estimates are used to check for collinearity between effects. Near collinearity may hamper the proper parameter estimation which is reflected by large standard errors (Ripley, Snijders and Preciado, 2010). For the here presented case, correlations between parameters are overall rather low. The highest value is -0.589 for the correlation between the cooperative experience and the absorptive capacity effect. However, both effects are significant (in model 2) which means that both effects should remain in the model. The random noise is high but the signal is strong enough that it exceeds the noise.

## 9.8    Conclusions

Competitive pressure forces firms to continuously develop new ideas, invent new technologies and bring new products to the market in order to survive the destructive part of Schumpeterian innovation competition. This holds particularly for the automotive industry in Germany, challenged by firms from emerging markets which are able to offer their products for lower prices. In the competition for new technological solutions, competences and cutting-edge knowledge are success factors. New knowledge stimulates the emergence of new ideas that can be transformed into innovation. Such knowledge can partly be generated internally in the companies' R&D laboratories. However, relying on internal knowledge generation is no longer sufficient. Participation in innovation networks which allow for access to external knowledge, and applying innovation cooperation as a strategic tool to acquire necessary knowledge which cannot be developed in-house opens up rich opportunities to complement and recombine the own knowledge-base. Thus, knowledge becomes the most important source of competitive advantage.

A first analytical aim of chapter 9 was to explore the structure of the automotive sample knowledge-base. I delineated a methodology which allows us analyzing a knowledge-base as a network of interrelated knowledge elements that are linked by patents. The network perspective makes it possible to have a deeper look into the knowledge structure of an organization or industry. The tie distribution shows a highly skewed picture which indicates that there are some knowledge elements which have a very large number of links providing them with the status of very important knowledge for the sample. A feature found in many networks, whether it be social, biological or physical networks, is a strong tendency for small-world properties. The positive results of the conducted small-world test shows that this feature can also be found in knowledge networks. Furthermore, the analysis of the importance of technologies over time provides an idea about technologies which become more or less important. Analyzing the e-mobility related knowledge confirms other studies by demonstrating that relatively little effort has been spent during the observation period to bring forward technologies for e-mobility. In a further step, the analysis was extended by the inclusion of the tie strength. Among the most central nodes in the sample we see only small fluctuations of the tie strength across the observations. Based on the data for tie strength, in analogy with Granovetter's (1973) theory of "the strength of weak ties", the hypothesis was derived that the ego networks of nodes that are linked by a strong tie have a strong tendency for overlapping due to knowledge relatedness as well as similarity features. A QAP correlation analysis shows that there is indeed a strong positive correlation between strong ties and network overlaps.

Firms access knowledge via their network relations. For the network composed of publicly funded R&D projects in the German automotive industry, structural as well as individual and dyadic covariates are shown to be relevant drivers of evolutionary change. In particular, the following main results were obtained: The establishment of cliques plays an important role in the evolution of innovation networks and the formation of triadic structures can be widely observed. The factors emphasized in the literature such as geographical distance, technological distance and cooperation experience are confirmed and hence explain network evolution of the sample firms in the German automotive industry. Also, firms with high levels of absorptive capacity tend to be more often involved in the investigated networks. The preference for modular knowledge-bases related to the automotive product architecture and manufacturing was not significant in the data. This needs further investigation since once R&D is increasingly shifted to suppliers and once the industry structures change their character from a strongly hierarchical architecture towards a more horizontal network organization, knowledge base structures might co-evolve. In fact, automotive suppliers are expected to gain even larger shares in the value chain during the next years. This tendency concerns production but also R&D. For R&D the share of the OEMs is expected to drop from 60 % in 2012 to only 47 % in 2025. Beneficiaries are suppliers and engineering service providers. This trend gets accelerated by the paradigmatic move in the power train towards electric engines (Oliver Wyman, 2012). In a nutshell, even though the OEMs have still the lead in product architecture design which requires sound knowledge across all relevant technologies, the trend clearly signifies that lower tier suppliers play a stronger role not only in production but particularly in R&D.

## 10.  Discussion and Further Research Avenues

Neo-Schumpeterian scholars suggest that we focus on the analysis of the knowledge-bases of firms and their role for entrepreneurial activities and managerial decisions if we want to understand patterns of industry change. Thus, the knowledge-based view of the firm paves the ground for the analysis of innovation network evolution. Evolutionary theory is very rich and offers a broad variety of concepts which have the potential for being applied in an (innovation) economic framework. Competitive pressure forces firms to continuously develop new ideas, invent new technologies and bring new products to the market in order to prevail on the field of creative destruction. This holds for the automotive industry in Germany (but also elsewhere in Europe) that has become challenged by firms from emerging markets. New knowledge serves as a basis for new ideas that can be transformed into products at a later stage. This knowledge is partly generated internally. However, we have seen that a more promising approach than solely relying on own R&D is to use networks as strategic tools to gain access to a broader variety of sources of knowledge which offer a multitude of possibilities to complement and recombine a firm's own knowledge-base.

Networks are evolving structures in terms or emerging and dissolving ties over time. A core research question was: What are the drivers and mechanisms that determine the change process? I applied a stochastic actor-based model which simulates network evolution between observation periods, explicitly models endogenous tie dependency and allows for the estimation of model parameters. For the networks that have been reconstructed from publicly funded R&D projects in the German automotive industry, structural as well as individual and dyadic covariates are relevant drivers: The formation of triadic structures could be observed; spatial proximity between firms increases the propensity to cooperate as well as experience (model 2); firms with high levels of absorptive capacity tend to be more often involved in networks. Internal R&D has not become obsolete but is a prerequisite to benefit from sources of collective knowledge.

This dissertation shows that evolutionary agent-based network models are valuable tools for improving our understanding about complex interaction structures of innovation networks. A particular feature is their ability to capture endogenous and exogenous driving forces simultaneously. Agent-based simulation models deliver new insights into network dynamics and have a great potential for even more sophisticated analyses in the future. They are used as a tool to disclose (plausible) causal relationships between firm strategy and the emerging network. The analysis of network evolution serves as an instrument to test the empirical relevance of attachment and dissolution mechanisms which are implemented in the artificial world of an agent-based simulation. On the other side, agent-based simulations can be used to develop additional hypotheses with respect to drivers of network evolution and to discover mechanisms which have not been theorized and tested yet. In this sense, the empirical research complements agent-based simulation models of an artificial world and vice versa.

A currently emerging research field concerns the development of co-evolutionary models. Most conventional economic models have a static character while reality exhibits dynamic features. Static means that we seek to explain a phenomenon (dependent variable) by a number of hypothesized factors and controls (independent variables). For instance, the question is asked: Are more central actors in a network more successful? Typically, this is a question about the linearity of causes and effects. While it is acknowledged that there might be a reverse causality problem, reverse causality gets hardly simultaneously modeled. However, in reality the direction of causality is often not clear and may go in both ways at the same time. The presented automotive network model captures network change over time and explains the driving forces. This approach allows not only for modeling tie changes but also takes changes in actor properties into account. A further step would be to explain these changes in actor properties with changes in network structures, e.g. the size of a firm's knowledge-base might be influenced by a firm's degree of network centrality. Moreover, as a function of the degree of network embeddedness, the innovativeness of a firm (measured by the patent output) may change. Firms which have more ties are more central, have better access to knowledge and more learning opportunities which allow them to be more innovative, build a larger knowledge-base and apply for more patents. That is, co-evolutionary models attempt to capture both-way-effects simultaneously. Thus, we may find that centrality not only fosters innovative success but also that the most successful actors become more central. In other words, the actor properties shape the network but the network may also shape actor properties. Moreover, we can think of co-evolutionary effects with regard to two types of networks which exert mutual influence. For instance, the production network must not be necessarily the same as the innovation network, but they mutually determine to some extend their evolutionary pattern. In addition, selection of cooperation partners based on firm preferences was the underlying basic driving force in this study. That is, selection of partners shapes the network pattern, as – for instance – firms make decisions in accordance with the homophily principle and select partners that are similar (or dissimilar) in one or more characteristics. Alternatively, there could be some co-evolutionary mechanism in place which stresses social influence. Firms may adapt their characteristics and match them with their partner's characteristics (Friedkin and Johnsen, 1999). In the context of innovation networks, this effect might lead to changes in firm characteristics which in turn influence the selection pattern. As a consequence of the high level of uncertainty which is inherent to innovation processes, firms may adapt their R&D expenses to the level of expenses of their partners leading to overall similar expenses. Another possibility is that a firm reallocates its resources in a certain technology field, for instance to hybrid powertrains, because their partners do the same. Alternatively, firm might do benchmarking (best practice) with regard to R&D expenses. Consequently, they may adapt their R&D expenses to the levels of their best performing partners. Thus, actor-based models allow for enhancing our knowledge not only about the processes of variety creation and selection, but also about the co-evolution of the agents within a network.

A couple of further questions may open interesting future research avenues: First, the hypothesized drivers have been tested for interfirm networks. It might be interesting to see if they are also valid for inventor networks or networks which involve research organizations. Second, the operationalisation of the tested effects could be subject to further inquiries. As for the case of the technological distance, there are sometimes various possibilities which may lead to diverging results. Third, other factors might be relevant but have not yet been tested such as the complementarity of knowledge-bases, the cultural distance between actors, which is particularly relevant for international cooperations, or institutional distances which matter in networks comprising public as well as private actors. Moreover, additional effects could be non-linear effects (non-linear functions). Fourth, with regard to the model as such, more sophisticated goodness-of-fit tests are currently developed and can help to increase the fit of the observations with regard to certain parameters such as the degree distribution. The applied test for time heterogeneity is a first step in this direction. Fifth, while some network effects seem to have a universal character, e.g. transitivity, other effects are specific for a certain industry. Consequently, effects should be tested in more industries. Estimations might be particularly interesting in industries from which we expect diverging results. For instance, Buchmann et al. (2014) study differences in network drivers between the automotive and the laser industry. Sixth, while I observe small-world characteristics for both, the interfirm network and the knowledge network, the causalities for the emergence of small-world networks require further investigations. Is there a natural tendency or are there rather social or technological reasons for their emergence?

From the presented analysis, a number of implications for managers and public policy actors can be derived: Since cooperation is a vital tool to access sources of relevant knowledge, location matters in the sense that there is a preference for a selection of partners which are spatially co-located. Consequently, re-location or the opening of a subsidiary in a region where the respective knowledge is bound could be an option for a firm which seeks to gain access to a specific network or specific knowledge-base. Moreover, for strategic decisions that are related to innovation, location is a factor which should be taken into account. Access to networks of knowledge is neither frictionless nor free of costs. Indeed, it is a highly social process. The firm which wants to benefit from the process of collective knowledge generation has to bring something in. New knowledge, generated internally or absorbed from sources outside the network, serves as a ticket to enter a circle of firms and is expected to be shared with other firms in the network. Own research is also required to build up an absorptive capacity which allows a firm to understand and make use of the knowledge which is stored in the network. The identified preference to select partners that have a somewhat similar knowledge-base indicates that new knowledge should be related to the existing knowledge which makes it easier to be understood and processed. Reputation is reflected and valued by the experience and the propensity to form cohesive network subgroups (closed triads). The preference of firms to select more experienced firms shows that the cooperation track record matters. A longer list of cooperative projects demonstrates that a firm is a reliable partner, has developed cooperative capabilities and has a history of successful cooperation projects. The reputational effects of being a preferred partner gets amplified by the triadic structure which accelerates the flow and improves the quality of information about the reputation, and puts social pressure on the actors to behave well.

Policy makers have often special groups of actors (SMEs, firms from less developed regions, public research institutes, etc.) in mind when they design and implement programs for public innovation support. The provision of public funds for cooperative R&D as well as the installation of innovation networks which are – at least in their infant phase – managed and promoted by public authorities is a policy tool which becomes increasingly used. For the evaluation of this policy tool, the question should be asked whether the network ties are actually formed between those organizations that are expected to form ties. Typically, large and established firms have more available resources to be informed about supported projects as well as the experience to successfully apply for funds, to lobby or to pay consultants. In contrast, small firms are typically the envisaged target of innovation support initiatives. If policies which contain such incentives have already been put in place, the depicted conceptual framework helps to evaluate if such policies are effective in that they change the preference structures with regard to partner selection of networking firms. The analysis of the automotive network shows that firm size is not a significant factor for the cooperation partner selection. As such, the analyzed policy tool for innovation support can be regarded as non-discriminatory with regard to smaller firms.

And to close the circle, we have seen that the complex interaction of actors shapes the character and properties of a network structure. In a similar vein, the complex interaction of ingredients determines the character and taste of nicely cooked dishes. Compared to interaction patterns in natural sciences, social networks are even more challenging to analyze since the behavior of actors changes over time in parallel with actor properties. In most cases, we will not be able to identify a clear-cut stable pattern. While there may be phases that are rather static, innovation from inside and outside creates new (disruptive) impetus for evolutionary change. This dissertation contributes to the understanding of network evolution. To fully understand the complexity of evolutionary network structures (if this will ever be possible), more studies on its drivers are necessary.

## A. Appendix: Tables

**Table A. 1:** IPC Degree centrality measures.

| IPC | De-gree | Posi-tion | IPC | De-gree | Posi-tion | IPC | De-gree | Posi-tion | IPC | De-gree | Posi-tion | IPC | De-gree | Posi-tion |
|---|---|---|---|---|---|---|---|---|---|---|---|---|---|---|
| 1998-2002 | | | 1999-2003 | | | 2000-2004 | | | 2001-2005 | | | 2002-2006 | | |
| B60R | 178 | 1 | H01L | 179 | 1 | H01L | 168 | 1 | H01L | 160 | 1 | B60R | 149 | 1 |
| H01L | 174 | 2 | B60R | 166 | 2 | B60R | 163 | 2 | B60R | 158 | 2 | H01L | 145 | 2 |
| G01N | 129 | 3 | B29C | 131 | 3 | B60K | 130 | 3 | G01N | 126 | 3 | G01N | 114 | 3 |
| B29C | 126 | 4 | G01N | 128 | 4 | G01N | 128 | 4 | B60K | 123 | 4 | B60K | 113 | 4 |
| H05K | 122 | 5 | B60K | 121 | 5 | B29C | 125 | 5 | B29C | 121 | 5 | B29C | 112 | 5 |
| B60K | 119 | 6 | H01M | 120 | 6 | H01M | 120 | 6 | H05K | 109 | 6 | H05K | 111 | 6 |
| H01M | 115 | 7 | H05K | 118 | 7 | B62D | 116 | 7 | B62D | 108 | 7 | B62D | 94 | 7 |
| F02M | 113 | 8 | F02M | 113 | 8 | H05K | 110 | 8 | H01M | 107 | 8 | B60T | 88 | 8 |
| B62D | 112 | 9 | B62D | 112 | 9 | F02M | 109 | 9 | G06F | 98 | 9 | F02M | 87 | 9 |
| F16H | 109 | 10 | F16H | 103 | 10 | G05B | 104 | 10 | F02M | 97 | 10 | F16H | 87 | 9 |
| G05B | 109 | 10 | G05B | 102 | 11 | G06F | 102 | 11 | B60T | 93 | 11 | G06F | 85 | 10 |
| B01D | 105 | 11 | F02D | 99 | 12 | H02K | 101 | 12 | F02D | 92 | 12 | H02K | 84 | 11 |
| G06F | 98 | 12 | B01D | 97 | 13 | B60T | 98 | 13 | G05B | 91 | 13 | F16C | 83 | 12 |
| F02D | 97 | 13 | H02K | 97 | 13 | F02D | 98 | 13 | H02K | 91 | 13 | H01M | 83 | 12 |
| G01D | 96 | 14 | G01D | 96 | 14 | G01D | 97 | 14 | G01D | 89 | 14 | F02D | 82 | 13 |
| H02K | 94 | 15 | G06F | 96 | 14 | F16H | 96 | 15 | F16H | 87 | 15 | G01D | 80 | 14 |
| B60T | 93 | 16 | B60T | 94 | 15 | F16J | 93 | 16 | B01D | 79 | 16 | G05B | 79 | 15 |
| G02B | 93 | 16 | B01J | 92 | 16 | B01D | 90 | 17 | B32B | 79 | 16 | B32B | 77 | 16 |
| B01J | 92 | 17 | B32B | 88 | 17 | F16F | 86 | 18 | F16C | 79 | 16 | G01R | 77 | 16 |
| F16J | 90 | 18 | F16J | 88 | 17 | B32B | 82 | 19 | G01R | 79 | 16 | B23K | 76 | 17 |
| B32B | 88 | 19 | F16F | 82 | 18 | F16C | 81 | 20 | F16J | 78 | 17 | G01S | 76 | 17 |
| F16C | 87 | 20 | G01R | 82 | 18 | G01R | 81 | 20 | B23K | 77 | 18 | B01D | 75 | 18 |
| G01B | 86 | 21 | G02B | 82 | 18 | G01S | 81 | 20 | G01S | 77 | 18 | F01D | 75 | 18 |
| C04B | 82 | 22 | H01R | 82 | 18 | B01J | 80 | 21 | F16F | 76 | 19 | F16F | 75 | 18 |
| F01N | 81 | 23 | F01N | 80 | 19 | H01R | 80 | 21 | H01R | 76 | 19 | H05B | 72 | 19 |
| H04N | 81 | 23 | F16C | 80 | 19 | G06K | 78 | 22 | F16D | 75 | 20 | F16D | 71 | 20 |
| C23C | 80 | 24 | G01B | 80 | 19 | G01B | 77 | 23 | H05B | 75 | 20 | C23C | 70 | 21 |
| H01R | 80 | 24 | F16K | 79 | 20 | C23C | 75 | 24 | B60H | 72 | 21 | F16J | 70 | 21 |
| B60S | 79 | 25 | H04B | 79 | 20 | G02B | 75 | 24 | F01N | 72 | 21 | F16B | 69 | 22 |
| F16D | 79 | 25 | G01S | 78 | 21 | H05B | 75 | 24 | B01J | 71 | 22 | G02B | 69 | 22 |
| F16K | 78 | 26 | H01H | 78 | 21 | B23K | 74 | 25 | C23C | 71 | 22 | G07C | 69 | 22 |
| H04B | 78 | 26 | B60S | 77 | 22 | F01N | 74 | 25 | B60N | 68 | 23 | F01N | 68 | 23 |
| F16B | 77 | 27 | C04B | 77 | 22 | F16B | 74 | 25 | F02B | 68 | 23 | B60H | 67 | 24 |
| F16F | 77 | 27 | C23C | 77 | 22 | H01H | 74 | 25 | G02B | 68 | 23 | H04L | 67 | 24 |
| G01P | 77 | 27 | F16D | 77 | 22 | F16D | 73 | 26 | G01B | 67 | 24 | G05D | 65 | 25 |
| G01S | 77 | 27 | B23K | 76 | 23 | G01P | 73 | 26 | F16B | 66 | 25 | F16L | 64 | 26 |
| G06K | 77 | 27 | G06K | 76 | 23 | B60H | 72 | 27 | G01P | 66 | 25 | H01F | 64 | 26 |
| G01R | 76 | 28 | H04N | 76 | 23 | C04B | 72 | 27 | G05D | 66 | 25 | H01R | 64 | 26 |
| B23K | 74 | 29 | H05B | 76 | 23 | H04B | 71 | 28 | G07C | 66 | 25 | B23P | 63 | 27 |
| H01H | 74 | 29 | G01P | 75 | 24 | F16K | 70 | 29 | G06K | 65 | 26 | B01J | 62 | 28 |

| | | | | | | | | | | | | | | |
|---|---|---|---|---|---|---|---|---|---|---|---|---|---|---|
| G01M | 73 | 30 | F16B | 74 | 25 | B60S | 69 | 30 | H04B | 65 | 26 | B60N | 62 | 28 |
| H05B | 72 | 31 | H04L | 74 | 25 | C01B | 69 | 30 | C04B | 64 | 27 | B60Q | 62 | 28 |
| H04L | 71 | 32 | B60H | 73 | 26 | G05D | 69 | 30 | F01D | 64 | 27 | C04B | 62 | 28 |
| B60H | 70 | 33 | G01M | 72 | 27 | G07C | 68 | 31 | F16K | 64 | 27 | F02B | 62 | 28 |
| B60Q | 70 | 33 | B60Q | 71 | 28 | B60N | 67 | 32 | H01F | 64 | 27 | G06K | 62 | 28 |
| G01L | 69 | 34 | G05D | 70 | 29 | B60Q | 67 | 32 | H04L | 64 | 27 | F16K | 61 | 29 |
| G07C | 66 | 35 | G01L | 68 | 30 | G01L | 67 | 32 | B60Q | 63 | 28 | H04B | 61 | 29 |
| G08G | 65 | 36 | C01B | 67 | 31 | G01M | 67 | 32 | F16L | 62 | 29 | G01C | 60 | 30 |
| C01B | 64 | 37 | G08G | 66 | 32 | H04L | 67 | 32 | H01H | 62 | 29 | G01P | 60 | 30 |
| F02B | 63 | 38 | G07C | 65 | 33 | G08G | 66 | 33 | A61B | 61 | 30 | G01B | 58 | 31 |
| G05D | 63 | 38 | H01F | 65 | 33 | F02B | 64 | 34 | B23P | 61 | 30 | A61B | 57 | 32 |
| A61B | 62 | 39 | F02B | 64 | 34 | H01F | 64 | 34 | G08G | 61 | 30 | G01L | 56 | 33 |
| B23P | 62 | 39 | A61B | 63 | 35 | F16L | 63 | 35 | B60S | 59 | 31 | G01M | 56 | 33 |
| B60N | 62 | 39 | B23P | 63 | 35 | H04N | 62 | 36 | G01L | 59 | 31 | H01H | 56 | 33 |
| H03K | 62 | 39 | B60N | 61 | 36 | A61B | 61 | 37 | H04N | 59 | 31 | G08G | 55 | 34 |
| G08C | 60 | 40 | F16L | 59 | 37 | F01D | 61 | 37 | C01B | 58 | 32 | B21D | 51 | 35 |
| H01F | 60 | 40 | H04M | 58 | 38 | G01C | 59 | 38 | G01C | 58 | 32 | B60J | 49 | 36 |
| B60J | 59 | 41 | G08C | 57 | 39 | B23P | 58 | 39 | B60G | 55 | 33 | B60G | 47 | 37 |
| G01C | 59 | 41 | F28F | 56 | 40 | B60G | 57 | 40 | G06T | 54 | 34 | C01B | 47 | 37 |
| H04M | 59 | 41 | B60G | 55 | 41 | G06T | 57 | 40 | H03K | 54 | 34 | H03K | 47 | 37 |
| G06T | 57 | 42 | B60J | 55 | 41 | B60J | 56 | 41 | B21D | 53 | 35 | H04N | 47 | 37 |
| B60W | 56 | 43 | G01C | 55 | 41 | F28F | 56 | 41 | B60J | 51 | 36 | B60S | 46 | 38 |
| H04Q | 56 | 43 | G06T | 55 | 41 | C09K | 55 | 42 | C09K | 51 | 36 | C09K | 46 | 38 |
| F16L | 54 | 44 | H03K | 55 | 41 | G08C | 54 | 43 | G01M | 51 | 36 | F28F | 45 | 39 |
| B60G | 53 | 45 | B60W | 54 | 42 | F15B | 53 | 44 | F15B | 50 | 37 | H02P | 45 | 39 |
| E05B | 53 | 45 | B60L | 53 | 43 | B21D | 52 | 45 | F28F | 49 | 38 | F21V | 44 | 40 |
| C22C | 52 | 46 | C08J | 53 | 43 | B60L | 52 | 45 | H01J | 48 | 39 | F28D | 44 | 40 |
| F15B | 52 | 46 | F01D | 53 | 43 | H03K | 52 | 45 | H02P | 47 | 40 | H04M | 44 | 40 |
| F28F | 52 | 46 | H03M | 53 | 43 | H04M | 52 | 45 | F21V | 46 | 41 | B23Q | 43 | 41 |
| B21D | 51 | 47 | B21D | 52 | 44 | B60C | 51 | 46 | H04M | 46 | 41 | F15B | 43 | 41 |
| C08J | 51 | 47 | C09K | 51 | 45 | B60W | 51 | 46 | B60C | 45 | 42 | G06T | 43 | 41 |
| C08L | 51 | 47 | F15B | 51 | 45 | F28D | 51 | 46 | F01P | 45 | 42 | G08B | 42 | 42 |
| H01Q | 51 | 47 | H01J | 51 | 45 | H01J | 51 | 46 | F28D | 45 | 42 | H02H | 42 | 42 |
| B29L | 50 | 48 | H01Q | 51 | 45 | F04D | 49 | 47 | G01F | 45 | 42 | H02J | 42 | 42 |
| B60L | 50 | 49 | H04Q | 50 | 46 | H03M | 49 | 47 | H01B | 45 | 42 | B60C | 41 | 43 |
| G08B | 50 | 49 | B29L | 49 | 47 | F01P | 48 | 48 | B60L | 44 | 43 | E05B | 41 | 43 |
| C09K | 49 | 50 | B60C | 48 | 48 | F21V | 48 | 48 | B60W | 44 | 43 | H01J | 41 | 43 |
| H01J | 49 | 50 | B05D | 47 | 49 | H01B | 48 | 48 | H02H | 44 | 43 | C08L | 40 | 44 |
| H02J | 49 | 50 | C22C | 47 | 49 | B05D | 47 | 49 | C08L | 43 | 44 | F02C | 40 | 44 |
| H03M | 49 | 50 | E05B | 47 | 49 | G01F | 46 | 50 | C22C | 43 | 44 | B05D | 39 | 45 |
| H02P | 48 | 51 | G08B | 47 | 49 | H01Q | 46 | 50 | F04B | 43 | 44 | B60W | 39 | 45 |
| D04H | 47 | 52 | H03H | 47 | 49 | H02P | 46 | 50 | F04D | 43 | 44 | C09D | 39 | 45 |
| F01D | 47 | 52 | B24B | 46 | 50 | E05B | 45 | 51 | G08B | 43 | 44 | F01L | 39 | 45 |
| G10K | 47 | 52 | D04H | 46 | 50 | G08B | 45 | 51 | G08C | 42 | 45 | F01P | 39 | 45 |
| B05D | 46 | 53 | F01P | 46 | 50 | C08J | 44 | 52 | B23Q | 41 | 46 | H01B | 39 | 45 |
| B22D | 46 | 53 | G09F | 46 | 50 | C08L | 44 | 52 | F02C | 41 | 46 | B24B | 38 | 46 |
| G06Q | 46 | 53 | H02J | 46 | 50 | C09D | 44 | 52 | H02J | 41 | 46 | E05F | 38 | 46 |

| | | | | | | | | | | | | | | |
|---|---|---|---|---|---|---|---|---|---|---|---|---|---|---|
| G11B | 46 | 53 | H02P | 46 | 50 | F01L | 44 | 52 | B01F | 40 | 47 | G01F | 38 | 46 |
| H03H | 46 | 53 | F04D | 45 | 51 | H03H | 44 | 52 | E05B | 40 | 47 | F04B | 37 | 47 |
| B60C | 45 | 54 | G06Q | 45 | 51 | H04Q | 44 | 52 | H03M | 40 | 47 | F04D | 37 | 47 |
| F01P | 45 | 54 | B65G | 44 | 52 | B24B | 43 | 53 | B05D | 39 | 48 | G09F | 37 | 47 |
| F02C | 45 | 54 | B23Q | 43 | 53 | B29L | 43 | 53 | G09F | 39 | 48 | B01F | 36 | 48 |
| F04D | 45 | 54 | C08K | 43 | 53 | C22C | 43 | 53 | B24B | 38 | 49 | H03M | 36 | 48 |
| G02F | 45 | 54 | C08L | 43 | 53 | G09F | 43 | 53 | F01L | 38 | 49 | B23D | 35 | 49 |
| B24B | 44 | 55 | C09D | 43 | 53 | H02J | 43 | 53 | G01K | 37 | 50 | B60L | 35 | 49 |
| C08K | 44 | 55 | F04B | 43 | 53 | B23Q | 42 | 54 | B22D | 36 | 51 | B65G | 35 | 49 |
| G01K | 44 | 55 | F28D | 43 | 53 | B65D | 42 | 54 | C08J | 36 | 51 | C08J | 35 | 49 |
| H01B | 44 | 55 | G01F | 43 | 53 | C08K | 42 | 54 | C08K | 36 | 51 | G08C | 35 | 49 |
| B23Q | 43 | 56 | H01B | 43 | 53 | G06Q | 42 | 54 | B25J | 35 | 52 | C08K | 34 | 50 |
| F28D | 43 | 56 | B22D | 42 | 54 | B65G | 41 | 55 | B29L | 35 | 52 | C22C | 34 | 50 |
| G03F | 43 | 56 | G01K | 42 | 54 | F02C | 41 | 55 | C09D | 35 | 52 | G01H | 33 | 51 |
| G09F | 43 | 56 | G10L | 42 | 54 | F04B | 41 | 55 | G01H | 35 | 52 | G01J | 33 | 51 |
| H02H | 43 | 56 | G11B | 42 | 54 | G01K | 40 | 56 | H01Q | 35 | 52 | G01K | 33 | 51 |
| B65D | 42 | 57 | F01L | 41 | 55 | H02H | 40 | 56 | H02M | 35 | 52 | H02M | 33 | 51 |
| F04B | 42 | 57 | F21V | 41 | 55 | B01F | 39 | 57 | B65G | 34 | 53 | C21D | 32 | 52 |
| G01F | 42 | 57 | G03F | 41 | 55 | H02M | 39 | 57 | E05F | 34 | 53 | F21S | 32 | 52 |
| H01G | 42 | 57 | H01G | 41 | 55 | B22D | 38 | 58 | F02F | 33 | 54 | G10L | 32 | 52 |
| B65G | 41 | 58 | H02H | 41 | 55 | D04H | 38 | 58 | F02N | 33 | 54 | B25B | 31 | 53 |
| C09D | 41 | 58 | B29D | 40 | 56 | G10K | 38 | 58 | F02P | 33 | 54 | B25F | 31 | 53 |
| F01L | 41 | 58 | B65D | 40 | 56 | G10L | 38 | 58 | F21S | 33 | 54 | B25J | 31 | 53 |
| F02F | 41 | 58 | F02C | 40 | 56 | B29D | 37 | 59 | G01J | 33 | 54 | B65D | 31 | 53 |
| F02P | 41 | 58 | B01F | 39 | 57 | F01M | 37 | 59 | G06Q | 33 | 54 | F01M | 31 | 53 |
| B29D | 39 | 59 | C08G | 39 | 57 | F21S | 37 | 59 | B22F | 32 | 55 | G06Q | 30 | 54 |
| B29K | 39 | 59 | F02P | 39 | 57 | F02F | 36 | 60 | B29D | 32 | 55 | H04Q | 30 | 54 |
| C08G | 39 | 59 | G10K | 39 | 57 | F02N | 36 | 60 | G10K | 32 | 55 | F02N | 29 | 55 |
| G10L | 39 | 59 | F02F | 38 | 58 | F02P | 36 | 60 | G10L | 32 | 55 | H01Q | 29 | 55 |
| F21V | 37 | 59 | H01C | 38 | 58 | G01J | 36 | 60 | H04Q | 32 | 55 | H04W | 29 | 55 |
| H02M | 37 | 59 | C21D | 37 | 59 | C21D | 35 | 61 | B65D | 31 | 56 | C25D | 28 | 56 |
| G03B | 36 | 60 | G02F | 37 | 59 | D06M | 35 | 61 | C21D | 31 | 56 | F02P | 28 | 56 |
| H01C | 36 | 60 | B29K | 35 | 60 | G02F | 35 | 61 | D06M | 31 | 56 | H04R | 28 | 56 |
| E05F | 35 | 61 | E05F | 35 | 60 | G03F | 35 | 61 | F01M | 31 | 56 | B22D | 27 | 57 |
| F25B | 35 | 61 | F21S | 35 | 60 | C08G | 34 | 62 | F25B | 30 | 57 | F02F | 27 | 57 |
| G05G | 35 | 61 | F25B | 35 | 60 | H01G | 34 | 62 | G02F | 30 | 57 | G02F | 27 | 57 |
| G11C | 35 | 61 | H02M | 35 | 60 | B22F | 33 | 63 | H03H | 30 | 57 | G10K | 27 | 57 |
| B25J | 34 | 62 | B22F | 34 | 61 | B25J | 33 | 63 | H04W | 30 | 57 | H03H | 27 | 57 |
| C08F | 34 | 62 | F01M | 34 | 61 | E05F | 33 | 63 | C25D | 29 | 58 | B23B | 26 | 58 |
| C21D | 34 | 62 | H02N | 34 | 61 | F24F | 33 | 63 | G03B | 29 | 58 | H02G | 26 | 58 |
| F02N | 34 | 62 | F02N | 33 | 62 | G01H | 33 | 63 | H04R | 29 | 58 | B21K | 25 | 59 |
| H02N | 34 | 62 | F24F | 33 | 62 | G03B | 32 | 64 | B21K | 28 | 59 | B29D | 25 | 59 |
| H04R | 34 | 62 | G03B | 33 | 62 | F25B | 31 | 65 | B25F | 28 | 59 | F25B | 25 | 59 |
| B01F | 33 | 63 | H04W | 33 | 62 | G11B | 31 | 65 | D04H | 28 | 59 | G03B | 25 | 59 |
| B22F | 33 | 63 | G01H | 32 | 63 | H04W | 31 | 65 | G11C | 28 | 59 | G09G | 25 | 59 |
| B81B | 33 | 63 | C25D | 31 | 64 | G05G | 30 | 66 | H01G | 28 | 59 | B21C | 24 | 60 |

| | | | | | | | | | | | | | | |
|---|---|---|---|---|---|---|---|---|---|---|---|---|---|---|
| F01M | 33 | 63 | D06M | 31 | 64 | H04R | 30 | 66 | H02G | 28 | 59 | B22F | 24 | 60 |
| G01V | 33 | 63 | G01J | 31 | 64 | B25F | 29 | 67 | B29K | 27 | 60 | B41M | 24 | 60 |
| F21S | 32 | 64 | G09G | 31 | 64 | B81B | 29 | 67 | G01V | 27 | 60 | D06M | 24 | 60 |
| F24F | 32 | 64 | B25J | 30 | 65 | C25D | 29 | 67 | B23D | 26 | 61 | F23Q | 24 | 60 |
| G09B | 32 | 64 | B61D | 30 | 65 | G09G | 29 | 67 | B25B | 26 | 61 | G01V | 24 | 60 |
| B65B | 30 | 65 | B61L | 30 | 65 | G11C | 29 | 67 | B61D | 26 | 61 | B05B | 23 | 61 |
| F01K | 30 | 65 | G05G | 30 | 65 | B21K | 28 | 68 | F23D | 26 | 61 | B21J | 23 | 61 |
| G01H | 30 | 65 | G11C | 30 | 65 | B23B | 28 | 68 | F23Q | 26 | 61 | C08G | 23 | 61 |
| G01J | 30 | 65 | H04R | 30 | 65 | G01V | 28 | 68 | F24F | 26 | 61 | D04H | 23 | 61 |
| G09G | 30 | 65 | B81B | 28 | 66 | H01C | 28 | 68 | G05G | 25 | 62 | B29K | 22 | 62 |
| H04W | 30 | 65 | C09J | 28 | 66 | H02N | 28 | 68 | G11B | 25 | 62 | F23D | 22 | 62 |
| A61L | 29 | 66 | F23D | 28 | 66 | B29K | 27 | 69 | B21C | 24 | 63 | G11C | 22 | 62 |
| B61D | 29 | 66 | A61L | 27 | 67 | B61D | 27 | 69 | B23B | 24 | 63 | H01C | 22 | 62 |
| B81C | 29 | 66 | B08B | 27 | 67 | F01K | 27 | 69 | B65B | 24 | 63 | H01G | 22 | 62 |
| H04J | 29 | 66 | B21K | 27 | 67 | F23M | 27 | 69 | B81B | 24 | 63 | B08B | 21 | 63 |
| A41D | 28 | 67 | B25F | 27 | 67 | B21C | 26 | 70 | F01K | 24 | 63 | B81B | 21 | 63 |
| B21K | 28 | 67 | B65B | 27 | 67 | B25B | 26 | 70 | F23M | 24 | 63 | C09J | 21 | 63 |
| B28B | 28 | 67 | G01V | 27 | 67 | F23D | 26 | 70 | G09G | 24 | 63 | F16G | 21 | 63 |
| B61L | 28 | 67 | G09B | 27 | 67 | H02G | 26 | 70 | H03F | 24 | 63 | F24F | 21 | 63 |
| C09J | 28 | 67 | H03F | 27 | 67 | A61L | 25 | 71 | C08G | 23 | 64 | F27D | 21 | 63 |
| C25D | 28 | 67 | H03L | 27 | 67 | B08B | 25 | 71 | F27D | 23 | 64 | G11B | 21 | 63 |
| D06M | 28 | 67 | A41D | 26 | 68 | B61L | 25 | 71 | G01G | 23 | 64 | B60B | 20 | 64 |
| B08B | 27 | 68 | B21C | 26 | 68 | B65H | 25 | 71 | H03G | 23 | 64 | B61D | 20 | 64 |
| C03C | 27 | 68 | C08F | 26 | 68 | F21Y | 25 | 71 | A61L | 22 | 65 | C22B | 20 | 64 |
| F23D | 27 | 68 | F01K | 26 | 68 | H03F | 25 | 71 | B41M | 22 | 65 | F23R | 20 | 64 |
| H02G | 27 | 68 | H02G | 26 | 68 | H04J | 25 | 71 | C09J | 22 | 65 | F25D | 20 | 64 |
| B65H | 26 | 69 | H04H | 26 | 68 | B65B | 24 | 72 | C22F | 22 | 65 | G01G | 20 | 64 |
| D01F | 26 | 69 | H04J | 26 | 68 | C09J | 24 | 72 | D06N | 22 | 65 | H03G | 20 | 64 |
| H03F | 26 | 69 | F21Y | 25 | 69 | G01G | 24 | 72 | H01C | 22 | 65 | B29L | 19 | 65 |
| H03L | 26 | 69 | G01G | 25 | 69 | B28B | 23 | 73 | B08B | 21 | 66 | B61L | 19 | 65 |
| H04H | 26 | 69 | B21B | 24 | 70 | B41M | 23 | 73 | D03D | 21 | 66 | F23N | 19 | 65 |
| B21B | 25 | 70 | B81C | 23 | 71 | C03C | 23 | 73 | F21Y | 21 | 66 | G05G | 19 | 65 |
| B29B | 25 | 70 | D03D | 23 | 71 | C22B | 23 | 73 | H02N | 21 | 66 | H01S | 19 | 65 |
| B21C | 24 | 71 | F04C | 23 | 71 | C23F | 23 | 73 | H04J | 21 | 66 | A61F | 18 | 66 |
| F21Y | 24 | 71 | F23K | 23 | 71 | D06N | 23 | 73 | A41D | 20 | 67 | C03C | 18 | 66 |
| B25F | 23 | 72 | H03G | 23 | 71 | F27D | 23 | 73 | F17C | 20 | 67 | C22F | 18 | 66 |
| F04C | 23 | 72 | B23B | 22 | 72 | A41D | 22 | 74 | G09B | 20 | 67 | F04C | 18 | 66 |
| F23M | 23 | 72 | B25B | 22 | 72 | B05B | 22 | 74 | H03L | 20 | 67 | F23M | 18 | 66 |
| H03G | 23 | 72 | B28B | 22 | 72 | B23D | 22 | 74 | H04H | 20 | 67 | G09B | 18 | 66 |
| H03J | 23 | 72 | B29B | 22 | 72 | F17C | 22 | 74 | B01L | 19 | 68 | H03F | 18 | 66 |
| C03B | 22 | 73 | B65H | 22 | 72 | F23Q | 22 | 74 | B05B | 19 | 68 | A41D | 17 | 67 |
| F23C | 22 | 73 | C03C | 22 | 72 | H03G | 22 | 74 | B28B | 19 | 68 | A61N | 17 | 67 |
| F23K | 22 | 73 | F17C | 22 | 72 | H03J | 22 | 74 | B60B | 19 | 68 | B01L | 17 | 67 |
| G01G | 22 | 73 | F23Q | 22 | 72 | H03L | 22 | 74 | B61L | 19 | 68 | B65B | 17 | 67 |
| G05F | 22 | 73 | F27D | 22 | 72 | C22F | 21 | 75 | B62J | 19 | 68 | B65H | 17 | 67 |
| G07F | 22 | 73 | G07F | 22 | 72 | D03D | 21 | 75 | B65H | 19 | 68 | F17C | 17 | 67 |
| H01T | 22 | 73 | H05H | 22 | 72 | F04C | 21 | 75 | E06B | 19 | 68 | F21Y | 17 | 67 |

| | | | | | | | | | | | | | | |
|---|---|---|---|---|---|---|---|---|---|---|---|---|---|---|
| B05B | 21 | 74 | B01L | 21 | 73 | F25D | 21 | 75 | F21W | 19 | 68 | H04H | 17 | 67 |
| B23B | 21 | 74 | B05B | 21 | 73 | G07B | 21 | 75 | F23R | 19 | 68 | H05G | 17 | 67 |
| F02G | 21 | 74 | B22C | 21 | 73 | G07F | 21 | 75 | H05G | 19 | 68 | B81C | 16 | 68 |
| H01P | 21 | 74 | B23D | 21 | 73 | H04H | 21 | 75 | B21B | 18 | 69 | C23F | 16 | 68 |
| H05H | 21 | 74 | D01F | 21 | 73 | B21B | 20 | 76 | B21J | 18 | 69 | C30B | 16 | 68 |
| A47C | 20 | 75 | F02G | 21 | 73 | G09B | 20 | 76 | B23C | 18 | 69 | D03D | 16 | 68 |
| B22C | 20 | 75 | F21W | 21 | 73 | B01L | 19 | 77 | B81C | 18 | 69 | F16M | 16 | 68 |
| B63H | 20 | 75 | F23M | 21 | 73 | B21H | 19 | 77 | C23F | 18 | 69 | F16N | 16 | 68 |
| F21W | 20 | 75 | H03J | 21 | 73 | B29B | 19 | 77 | F23N | 18 | 69 | F16P | 16 | 68 |
| F23N | 20 | 75 | B63H | 20 | 74 | B62J | 19 | 77 | F25D | 18 | 69 | F21K | 16 | 68 |
| F23Q | 20 | 75 | C03B | 20 | 74 | B63H | 19 | 77 | G01T | 18 | 69 | F23C | 16 | 68 |
| G06N | 20 | 75 | C22F | 20 | 74 | B81C | 19 | 77 | G03F | 18 | 69 | G05F | 16 | 68 |
| B21J | 19 | 76 | F23C | 20 | 74 | F21W | 19 | 77 | H01S | 18 | 69 | G21K | 16 | 68 |
| D06N | 19 | 76 | F25D | 20 | 74 | G01T | 19 | 77 | H05H | 18 | 69 | A47B | 15 | 69 |
| G01T | 19 | 76 | G06N | 20 | 74 | G05F | 19 | 77 | A47L | 17 | 70 | B21B | 15 | 69 |
| G07B | 19 | 76 | H01T | 20 | 74 | G06N | 19 | 77 | C02F | 17 | 70 | B60P | 15 | 69 |
| G21K | 19 | 76 | B03B | 19 | 75 | H05G | 19 | 77 | C03B | 17 | 70 | B63B | 15 | 69 |
| B01B | 18 | 77 | B41M | 19 | 75 | B03B | 18 | 78 | C03C | 17 | 70 | C02F | 15 | 69 |
| B09B | 18 | 77 | B60B | 19 | 75 | B22C | 18 | 78 | C22B | 17 | 70 | C25F | 15 | 69 |
| B21H | 18 | 77 | C23F | 19 | 75 | B63G | 18 | 78 | C30B | 17 | 70 | E05C | 15 | 69 |
| B60B | 18 | 77 | D06N | 19 | 75 | C03B | 18 | 78 | F04C | 17 | 70 | E06B | 15 | 69 |
| B63G | 18 | 77 | G05F | 19 | 75 | C25F | 18 | 78 | F16M | 17 | 70 | F01K | 15 | 69 |
| C02F | 18 | 77 | G07B | 19 | 75 | C30B | 18 | 78 | F21K | 17 | 70 | F24D | 15 | 69 |
| E01F | 18 | 77 | B01B | 18 | 76 | F23N | 18 | 78 | G05F | 17 | 70 | G01T | 15 | 69 |
| F25D | 18 | 77 | B09B | 18 | 76 | H01T | 18 | 78 | G21K | 17 | 70 | H01P | 15 | 69 |
| B01L | 17 | 78 | B21H | 18 | 76 | H05H | 18 | 78 | H01P | 17 | 70 | H03L | 15 | 69 |
| B67D | 17 | 78 | B26D | 18 | 76 | B09B | 17 | 79 | H03J | 17 | 70 | H04J | 15 | 69 |
| C22F | 17 | 78 | B44C | 18 | 76 | B21J | 17 | 79 | B21H | 16 | 71 | A47L | 14 | 70 |
| C23F | 17 | 78 | B63B | 18 | 76 | B23C | 17 | 79 | B24C | 16 | 71 | A61L | 14 | 70 |
| E06B | 17 | 78 | B63G | 18 | 76 | B60B | 17 | 79 | B60P | 16 | 71 | B24C | 14 | 70 |
| H03B | 17 | 78 | G01T | 18 | 76 | B64D | 17 | 79 | B63H | 16 | 71 | B29B | 14 | 70 |
| A61M | 16 | 79 | H01P | 18 | 76 | C02F | 17 | 79 | B64D | 16 | 71 | B44C | 14 | 70 |
| B03B | 16 | 79 | B21J | 17 | 77 | D01F | 17 | 79 | C23G | 16 | 71 | B67D | 14 | 70 |
| B07C | 16 | 79 | B64D | 17 | 77 | E06B | 17 | 79 | C25F | 16 | 71 | D06N | 14 | 70 |
| B42D | 16 | 79 | C30B | 17 | 77 | F02G | 17 | 79 | F02G | 16 | 71 | F24H | 14 | 70 |
| B44C | 16 | 79 | F01B | 17 | 77 | F21K | 17 | 79 | F16G | 16 | 71 | G07B | 14 | 70 |
| B63B | 16 | 79 | F23N | 17 | 77 | F23R | 17 | 79 | F23C | 16 | 71 | G07F | 14 | 70 |
| C07B | 16 | 79 | F23R | 17 | 77 | G21K | 17 | 79 | G07B | 16 | 71 | H03J | 14 | 70 |
| C12M | 16 | 79 | F24H | 17 | 77 | H01P | 17 | 79 | D01F | 15 | 72 | H05H | 14 | 70 |
| C12N | 16 | 79 | G21K | 17 | 77 | B01B | 16 | 80 | F16N | 15 | 72 | B05C | 13 | 71 |
| F16G | 16 | 79 | A47C | 16 | 78 | B24C | 16 | 80 | G07F | 15 | 72 | B23C | 13 | 71 |
| F23J | 16 | 79 | A61M | 16 | 78 | B44C | 16 | 80 | H01T | 15 | 72 | B62J | 13 | 71 |
| F24H | 16 | 79 | C12N | 16 | 78 | B60P | 16 | 80 | A01D | 14 | 73 | C07D | 13 | 71 |
| F27D | 16 | 79 | H03B | 16 | 78 | B63B | 16 | 80 | A61F | 14 | 73 | C07F | 13 | 71 |
| H04S | 16 | 79 | H05G | 16 | 78 | C07D | 16 | 80 | A61N | 14 | 73 | F01B | 13 | 71 |
| B05C | 15 | 80 | A47L | 15 | 79 | C23G | 16 | 80 | B26D | 14 | 73 | F21W | 13 | 71 |

| | | | | | | | | | | | | | | |
|---|---|---|---|---|---|---|---|---|---|---|---|---|---|---|
| B06B | 15 | 80 | B60P | 15 | 79 | F16G | 16 | 80 | B67D | 14 | 73 | F27B | 13 | 71 |
| B25B | 15 | 80 | C02F | 15 | 79 | F16M | 16 | 80 | C08F | 14 | 73 | G04B | 13 | 71 |
| B64D | 15 | 80 | C12M | 15 | 79 | F16N | 16 | 80 | F01B | 14 | 73 | H01K | 13 | 71 |
| D03D | 15 | 80 | C22B | 15 | 79 | F23K | 16 | 80 | F16P | 14 | 73 | H03B | 13 | 71 |
| F01B | 15 | 80 | E06B | 15 | 79 | H03B | 16 | 80 | F24D | 14 | 73 | H03B | 13 | 71 |
| F23L | 15 | 80 | F23L | 15 | 79 | A01B | 15 | 81 | F27B | 14 | 73 | B21H | 12 | 72 |
| F23R | 15 | 80 | B05C | 14 | 80 | A47C | 15 | 81 | G01W | 14 | 73 | B23H | 12 | 72 |
| H03D | 15 | 80 | B82B | 14 | 80 | A47L | 15 | 81 | H03B | 14 | 73 | B25D | 12 | 72 |
| H05G | 15 | 80 | C25B | 14 | 80 | B26D | 15 | 81 | B05C | 13 | 74 | B25H | 12 | 72 |
| A47L | 14 | 81 | C40B | 14 | 80 | B67D | 15 | 81 | B24D | 13 | 74 | B27B | 12 | 72 |
| B23D | 14 | 81 | F16G | 14 | 80 | C12N | 15 | 81 | B62K | 13 | 74 | B28B | 12 | 72 |
| B26D | 14 | 81 | F21K | 14 | 80 | C40B | 15 | 81 | C07D | 13 | 74 | C23G | 12 | 72 |
| B61F | 14 | 81 | G06G | 14 | 80 | F01B | 15 | 81 | C25B | 13 | 74 | C25B | 12 | 72 |
| B82B | 14 | 81 | H01S | 14 | 80 | F23C | 15 | 81 | E05C | 13 | 74 | E04B | 12 | 72 |
| C07D | 14 | 81 | H03D | 14 | 80 | B66F | 14 | 82 | E05D | 13 | 74 | E05D | 12 | 72 |
| C30B | 14 | 81 | A01B | 13 | 81 | B82B | 14 | 82 | F24H | 13 | 74 | F02G | 12 | 72 |
| E04F | 14 | 81 | B07C | 13 | 81 | C08F | 14 | 82 | A47C | 12 | 75 | G01W | 12 | 72 |
| F17C | 14 | 81 | B67D | 13 | 81 | C12M | 14 | 82 | A61H | 12 | 75 | G03F | 12 | 72 |
| F23G | 14 | 81 | C07B | 13 | 81 | E01F | 14 | 82 | A61K | 12 | 75 | G12B | 12 | 72 |
| F42B | 14 | 81 | C07D | 13 | 81 | F16P | 14 | 82 | B01B | 12 | 75 | H02N | 12 | 72 |
| A01B | 13 | 82 | C07F | 13 | 81 | F23L | 14 | 82 | B25H | 12 | 75 | H04S | 12 | 72 |
| A47B | 13 | 82 | C10L | 13 | 81 | G01W | 14 | 82 | B27B | 12 | 75 | A01D | 11 | 73 |
| D04B | 13 | 82 | C25F | 13 | 81 | G06G | 14 | 82 | B63B | 12 | 75 | B24D | 11 | 73 |
| F01C | 13 | 82 | E01F | 13 | 81 | H01S | 14 | 82 | B63J | 12 | 75 | B26B | 11 | 73 |
| F16M | 13 | 82 | E04F | 13 | 81 | A61F | 13 | 83 | C07F | 12 | 75 | B26D | 11 | 73 |
| H01S | 13 | 82 | F16P | 13 | 81 | A61M | 13 | 83 | H04S | 12 | 75 | B62K | 11 | 73 |
| B03C | 12 | 83 | F42B | 13 | 81 | B05C | 13 | 83 | A61J | 11 | 76 | B63H | 11 | 73 |
| B07B | 12 | 83 | H04S | 13 | 81 | B25H | 13 | 83 | A61M | 11 | 76 | C03B | 11 | 73 |
| B23C | 12 | 83 | A47B | 12 | 82 | B27B | 13 | 83 | B07C | 11 | 76 | C08F | 11 | 73 |
| B60P | 12 | 83 | A61K | 12 | 82 | C07B | 13 | 83 | B22C | 11 | 76 | C12Q | 11 | 73 |
| B61C | 12 | 83 | B03C | 12 | 82 | C25B | 13 | 83 | B25D | 11 | 76 | F03D | 11 | 73 |
| B66F | 12 | 83 | B23C | 12 | 82 | F24H | 13 | 83 | B29B | 11 | 76 | A47C | 10 | 74 |
| C07F | 12 | 83 | B24C | 12 | 82 | F27B | 13 | 83 | B44C | 11 | 76 | A61J | 10 | 74 |
| C10L | 12 | 83 | B27B | 12 | 82 | H01K | 13 | 83 | B61F | 11 | 76 | B26F | 10 | 74 |
| C23G | 12 | 83 | B61K | 12 | 82 | A01D | 12 | 84 | B64C | 11 | 76 | B27G | 10 | 74 |
| C40B | 12 | 83 | B66F | 12 | 82 | A61N | 12 | 84 | B66F | 11 | 76 | B41F | 10 | 74 |
| F16P | 12 | 83 | C12Q | 12 | 82 | B03C | 12 | 84 | F23L | 11 | 76 | B61F | 10 | 74 |
| F21K | 12 | 83 | C23G | 12 | 82 | B24D | 12 | 84 | F24J | 11 | 76 | C40B | 10 | 74 |
| G06G | 12 | 83 | F16M | 12 | 82 | C10L | 12 | 84 | G06N | 11 | 76 | F01C | 10 | 74 |
| G09C | 12 | 83 | F24J | 12 | 82 | C12Q | 12 | 84 | H01K | 11 | 76 | H01T | 10 | 74 |
| A61F | 11 | 84 | F27B | 12 | 82 | E05C | 12 | 84 | C25C | 10 | 77 | H03D | 10 | 74 |
| A61J | 11 | 84 | A61F | 11 | 83 | E05D | 12 | 84 | C40B | 10 | 77 | A43B | 9 | 75 |
| A61K | 11 | 84 | A61J | 11 | 83 | F01C | 12 | 84 | E04B | 10 | 77 | A61M | 9 | 75 |
| B25H | 11 | 84 | B06B | 11 | 83 | H04S | 12 | 84 | E21B | 10 | 77 | B01B | 9 | 75 |
| B41M | 11 | 84 | B24D | 11 | 83 | A61J | 11 | 85 | G06G | 10 | 77 | B30B | 9 | 75 |
| B61K | 11 | 84 | B42D | 11 | 83 | A61K | 11 | 85 | H03D | 10 | 77 | B60D | 9 | 75 |
| C07C | 11 | 84 | B60D | 11 | 83 | B07C | 11 | 85 | A61G | 9 | 78 | B64C | 9 | 75 |
| | | | | | | | | | | | | C25C | 9 | 75 |

| | | | | | | | | | | | | | | |
|---|---|---|---|---|---|---|---|---|---|---|---|---|---|---|
| C22B | 11 | 84 | B61F | 11 | 83 | B25D | 11 | 85 | B03D | 9 | 78 | E04H | 9 | 75 |
| E05D | 11 | 84 | B64C | 11 | 83 | C09C | 11 | 85 | B09B | 9 | 78 | E21B | 9 | 75 |
| F03D | 11 | 84 | C09C | 11 | 83 | F03D | 11 | 85 | B30B | 9 | 78 | F15D | 9 | 75 |
| F15D | 11 | 84 | E05D | 11 | 83 | F24D | 11 | 85 | B60D | 9 | 78 | F22B | 9 | 75 |
| F22B | 11 | 84 | F15D | 11 | 83 | H03D | 11 | 85 | B64F | 9 | 78 | F23K | 9 | 75 |
| H03C | 11 | 84 | F22B | 11 | 83 | A61H | 10 | 86 | C12Q | 9 | 78 | F24J | 9 | 75 |
| A61H | 10 | 85 | F23G | 11 | 83 | B07B | 10 | 86 | F01C | 9 | 78 | H02B | 9 | 75 |
| A61N | 10 | 85 | F23J | 11 | 83 | B60D | 10 | 86 | F04F | 9 | 78 | B07C | 8 | 76 |
| B60D | 10 | 85 | F24D | 11 | 83 | B61F | 10 | 86 | G04G | 9 | 78 | B22C | 8 | 76 |
| B62J | 10 | 85 | G09C | 11 | 83 | B62K | 10 | 86 | G12B | 9 | 78 | B60M | 8 | 76 |
| B63J | 10 | 85 | H01K | 11 | 83 | B63J | 10 | 86 | A43B | 8 | 79 | B61C | 8 | 76 |
| C09C | 10 | 85 | H04K | 11 | 83 | B64C | 10 | 86 | A47B | 8 | 79 | B64F | 8 | 76 |
| C10M | 10 | 85 | A43B | 10 | 84 | C25C | 10 | 86 | B03B | 8 | 79 | B66F | 8 | 76 |
| C12Q | 10 | 85 | A61H | 10 | 84 | E04B | 10 | 86 | B06B | 8 | 79 | B68G | 8 | 76 |
| E05C | 10 | 85 | A61N | 10 | 84 | E04F | 10 | 86 | B23H | 8 | 79 | C01C | 8 | 76 |
| F24J | 10 | 85 | B07B | 10 | 84 | G09C | 10 | 86 | B41F | 8 | 79 | C21B | 8 | 76 |
| F41H | 10 | 85 | B25D | 10 | 84 | H02B | 10 | 86 | B61C | 8 | 79 | E02F | 8 | 76 |
| G01W | 10 | 85 | B44F | 10 | 84 | H05F | 10 | 86 | B67B | 8 | 79 | F02K | 8 | 76 |
| G03G | 10 | 85 | B62K | 10 | 84 | B02C | 9 | 87 | B68G | 8 | 79 | F23L | 8 | 76 |
| H05F | 10 | 85 | B63J | 10 | 84 | B03D | 9 | 87 | B82B | 8 | 79 | G06N | 8 | 76 |
| B02C | 9 | 86 | C07C | 10 | 84 | B06B | 9 | 87 | C01C | 8 | 79 | A44B | 7 | 77 |
| B24C | 9 | 86 | C10M | 10 | 84 | B26F | 9 | 87 | C11D | 8 | 79 | B06B | 7 | 77 |
| B24D | 9 | 86 | H05F | 10 | 84 | B30B | 9 | 87 | D04B | 8 | 79 | B25G | 7 | 77 |
| B27B | 9 | 86 | A01D | 9 | 85 | B42D | 9 | 87 | F02K | 8 | 79 | B62B | 7 | 77 |
| B28D | 9 | 86 | B02C | 9 | 85 | B64F | 9 | 87 | F03D | 8 | 79 | B63G | 7 | 77 |
| B30B | 9 | 86 | B30B | 9 | 85 | E04H | 9 | 87 | F15D | 8 | 79 | B63J | 7 | 77 |
| B44F | 9 | 86 | B60M | 9 | 85 | E21B | 9 | 87 | F22B | 8 | 79 | C07C | 7 | 77 |
| B64C | 9 | 86 | B61C | 9 | 85 | F02K | 9 | 87 | F23K | 8 | 79 | D21F | 7 | 77 |
| C09B | 9 | 86 | B62J | 9 | 85 | F04F | 9 | 87 | H02B | 8 | 79 | G06G | 7 | 77 |
| C25F | 9 | 86 | C11D | 9 | 85 | G12B | 9 | 87 | H05F | 8 | 79 | A61C | 6 | 78 |
| E02F | 9 | 86 | C25C | 9 | 85 | A43B | 8 | 88 | B23F | 7 | 80 | A61G | 6 | 78 |
| E04B | 9 | 86 | E04B | 9 | 85 | A47B | 8 | 88 | B25G | 7 | 80 | A61H | 6 | 78 |
| E04C | 9 | 86 | E04H | 9 | 85 | B23H | 8 | 88 | B26F | 7 | 80 | A61K | 6 | 78 |
| E21B | 9 | 86 | E05C | 9 | 85 | B44F | 8 | 88 | B28D | 7 | 80 | B27C | 6 | 78 |
| F16N | 9 | 86 | E21B | 9 | 85 | B60M | 8 | 88 | B63G | 7 | 80 | B27D | 6 | 78 |
| F27B | 9 | 86 | F01C | 9 | 85 | B61C | 8 | 88 | C07C | 7 | 80 | B28D | 6 | 78 |
| G12B | 9 | 86 | F02K | 9 | 85 | B61K | 8 | 88 | C12M | 7 | 80 | B61K | 6 | 78 |
| H02B | 9 | 86 | F03D | 9 | 85 | B67B | 8 | 88 | C12N | 7 | 80 | B64D | 6 | 78 |
| H04K | 9 | 86 | F16N | 9 | 85 | C07H | 8 | 88 | D06C | 7 | 80 | C01G | 6 | 78 |
| A61P | 8 | 87 | G01W | 9 | 85 | C11D | 8 | 88 | D21F | 7 | 80 | C11D | 6 | 78 |
| B41N | 8 | 87 | G04G | 9 | 85 | D04B | 8 | 88 | E02F | 7 | 80 | D01F | 6 | 78 |
| B62K | 8 | 87 | G12B | 9 | 85 | E02F | 8 | 88 | E04F | 7 | 80 | D04B | 6 | 78 |
| B66C | 8 | 87 | A61P | 8 | 86 | F15D | 8 | 88 | F23G | 7 | 80 | D06C | 6 | 78 |
| B67B | 8 | 87 | B03D | 8 | 86 | F22B | 8 | 88 | A01B | 6 | 81 | D21C | 6 | 78 |
| C07H | 8 | 87 | B41N | 8 | 86 | F23J | 8 | 88 | A44B | 6 | 81 | D21H | 6 | 78 |
| F02K | 8 | 87 | B67B | 8 | 86 | F24J | 8 | 88 | A61C | 6 | 81 | E03F | 6 | 78 |

| | | | | | | | | | | | | | | |
|---|---|---|---|---|---|---|---|---|---|---|---|---|---|---|
| F03B | 8 | 87 | C07H | 8 | 86 | G04G | 8 | 88 | B27C | 6 | 81 | F03G | 6 | 78 |
| G04G | 8 | 87 | D04B | 8 | 86 | H03C | 8 | 88 | B27G | 6 | 81 | F04F | 6 | 78 |
| G07G | 8 | 87 | D06H | 8 | 86 | A45C | 7 | 89 | B60M | 6 | 81 | G07D | 6 | 78 |
| A43B | 7 | 88 | E02F | 8 | 86 | A61G | 7 | 89 | B62M | 6 | 81 | A01G | 5 | 79 |
| A47G | 7 | 88 | F03B | 8 | 86 | B23F | 7 | 89 | C01G | 6 | 81 | A47G | 5 | 79 |
| A61G | 7 | 88 | F04F | 8 | 86 | B25G | 7 | 89 | C07B | 6 | 81 | B03D | 5 | 79 |
| B25D | 7 | 88 | H03C | 8 | 86 | B26B | 7 | 89 | C09B | 6 | 81 | B41J | 5 | 79 |
| B27N | 7 | 88 | A44B | 7 | 87 | B28D | 7 | 89 | C10L | 6 | 81 | B42D | 5 | 79 |
| C25B | 7 | 88 | A45C | 7 | 87 | B41F | 7 | 89 | D02G | 6 | 81 | B62M | 5 | 79 |
| E04H | 7 | 88 | A61G | 7 | 87 | B68G | 7 | 89 | D21C | 6 | 81 | B66C | 5 | 79 |
| F03G | 7 | 88 | B25H | 7 | 87 | C07C | 7 | 89 | D21H | 6 | 81 | C09C | 5 | 79 |
| F04F | 7 | 88 | B26F | 7 | 87 | C07F | 7 | 89 | E01F | 6 | 81 | C14C | 5 | 79 |
| F15C | 7 | 88 | B28D | 7 | 87 | D06C | 7 | 89 | E03F | 6 | 81 | C21C | 5 | 79 |
| F28C | 7 | 88 | B64F | 7 | 87 | D06F | 7 | 89 | E04H | 6 | 81 | D06F | 5 | 79 |
| G03C | 7 | 88 | B66C | 7 | 87 | F23G | 7 | 89 | G07D | 6 | 81 | D21G | 5 | 79 |
| G03H | 7 | 88 | D06C | 7 | 87 | F28C | 7 | 89 | G09C | 6 | 81 | E01F | 5 | 79 |
| G04B | 7 | 88 | D06F | 7 | 87 | G03C | 7 | 89 | A63B | 5 | 82 | G03H | 5 | 79 |
| G04F | 7 | 88 | F03G | 7 | 87 | G03H | 7 | 89 | B41J | 5 | 82 | A01B | 4 | 80 |
| H01K | 7 | 88 | F28C | 7 | 87 | A44B | 6 | 90 | B42D | 5 | 82 | A62C | 4 | 80 |
| A41F | 6 | 89 | G03C | 7 | 87 | B27C | 6 | 90 | B66C | 5 | 82 | B03B | 4 | 80 |
| A44B | 6 | 89 | G03H | 7 | 87 | B27G | 6 | 90 | C09C | 5 | 82 | B03C | 4 | 80 |
| A45C | 6 | 89 | G07G | 7 | 87 | B41J | 6 | 90 | C14C | 5 | 82 | B04B | 4 | 80 |
| A63B | 6 | 89 | H02B | 7 | 87 | B62H | 6 | 90 | C21B | 5 | 82 | B21F | 4 | 80 |
| B27G | 6 | 89 | A41F | 6 | 88 | B66C | 6 | 90 | C21C | 5 | 82 | B23F | 4 | 80 |
| B41J | 6 | 89 | A47G | 6 | 88 | C09B | 6 | 90 | F42B | 5 | 82 | B28C | 4 | 80 |
| B60M | 6 | 89 | B26B | 6 | 88 | D06B | 6 | 90 | G03H | 5 | 82 | B61B | 4 | 80 |
| B62H | 6 | 89 | B27G | 6 | 88 | F03B | 6 | 90 | G04B | 5 | 82 | B66B | 4 | 80 |
| C01C | 6 | 89 | B41J | 6 | 88 | F42B | 6 | 90 | A47G | 4 | 83 | C06B | 4 | 80 |
| C01F | 6 | 89 | B62H | 6 | 88 | G03G | 6 | 90 | A61P | 4 | 83 | C06C | 4 | 80 |
| D06B | 6 | 89 | C01F | 6 | 88 | G07D | 6 | 90 | A63F | 4 | 83 | C07B | 4 | 80 |
| D06F | 6 | 89 | C09B | 6 | 88 | A61C | 5 | 91 | B03C | 4 | 83 | C09B | 4 | 80 |
| D07B | 6 | 89 | D06B | 6 | 88 | B21F | 5 | 91 | B04B | 4 | 83 | C10M | 4 | 80 |
| F24D | 6 | 89 | F41H | 6 | 88 | B28C | 5 | 91 | B21F | 4 | 83 | C12M | 4 | 80 |
| A01D | 5 | 90 | G03G | 6 | 88 | C01C | 5 | 91 | B26B | 4 | 83 | D02G | 4 | 80 |
| A23G | 5 | 90 | G07D | 6 | 88 | C10M | 5 | 91 | B28C | 4 | 83 | D06H | 4 | 80 |
| A61C | 5 | 90 | A61C | 5 | 89 | C14C | 5 | 91 | B61K | 4 | 83 | D07B | 4 | 80 |
| B04B | 5 | 90 | B04B | 5 | 89 | C21B | 5 | 91 | B62B | 4 | 83 | D21B | 4 | 80 |
| B26B | 5 | 90 | B41F | 5 | 89 | D05C | 5 | 91 | C06B | 4 | 83 | E01B | 4 | 80 |
| B26F | 5 | 90 | B61B | 5 | 89 | D21C | 5 | 91 | C06C | 4 | 83 | E04C | 4 | 80 |
| B61B | 5 | 90 | B61J | 5 | 89 | D21F | 5 | 91 | D06F | 4 | 83 | E04F | 4 | 80 |
| B61J | 5 | 90 | B66B | 5 | 89 | G01Q | 5 | 91 | D06H | 4 | 83 | E04G | 4 | 80 |
| B62B | 5 | 90 | C01C | 5 | 89 | G04B | 5 | 91 | D07B | 4 | 83 | F03C | 4 | 80 |
| B64F | 5 | 90 | C10G | 5 | 89 | H04K | 5 | 91 | D21B | 4 | 83 | F17D | 4 | 80 |
| B66B | 5 | 90 | C21B | 5 | 89 | A47G | 4 | 92 | E01B | 4 | 83 | F24C | 4 | 80 |
| B67C | 5 | 90 | D04D | 5 | 89 | A61P | 4 | 92 | E01C | 4 | 83 | F28C | 4 | 80 |
| C01G | 5 | 90 | D05C | 5 | 89 | A63B | 4 | 92 | E04C | 4 | 83 | F41H | 4 | 80 |
| C10B | 5 | 90 | D21C | 5 | 89 | B04B | 4 | 92 | E04G | 4 | 83 | F42B | 4 | 80 |

| | | | | | | | | | | | | | | |
|---|---|---|---|---|---|---|---|---|---|---|---|---|---|---|
| C10G | 5 | 90 | D21F | 5 | 89 | B04C | 4 | 92 | F03C | 4 | 83 | G03C | 4 | 80 |
| D02G | 5 | 90 | F15C | 5 | 89 | B27N | 4 | 92 | F15C | 4 | 83 | G03D | 4 | 80 |
| D04D | 5 | 90 | G01Q | 5 | 89 | B41N | 4 | 92 | F17D | 4 | 83 | A43C | 3 | 81 |
| D05B | 5 | 90 | G04B | 5 | 89 | B61B | 4 | 92 | F23J | 4 | 83 | A45C | 3 | 81 |
| D05C | 5 | 90 | A42B | 4 | 90 | C01F | 4 | 92 | F28C | 4 | 83 | A63B | 3 | 81 |
| D06C | 5 | 90 | A44C | 4 | 90 | C01G | 4 | 92 | F41H | 4 | 83 | A63F | 3 | 81 |
| D06H | 5 | 90 | A63B | 4 | 90 | C06B | 4 | 92 | G03C | 4 | 83 | B04C | 3 | 81 |
| G01Q | 5 | 90 | B04C | 4 | 90 | C06C | 4 | 92 | G03D | 4 | 83 | B09B | 3 | 81 |
| A42B | 4 | 91 | B23H | 4 | 90 | C09G | 4 | 92 | G04F | 4 | 83 | B21L | 3 | 81 |
| A44C | 4 | 91 | B27N | 4 | 90 | C12R | 4 | 92 | H03C | 4 | 83 | B41N | 3 | 81 |
| B03D | 4 | 91 | B61G | 4 | 90 | D02G | 4 | 92 | A43C | 3 | 84 | B61G | 3 | 81 |
| B04C | 4 | 91 | B67C | 4 | 90 | D06H | 4 | 92 | A45C | 3 | 84 | B82B | 3 | 81 |
| B23F | 4 | 91 | C01G | 4 | 90 | D21B | 4 | 92 | A62B | 3 | 84 | C06D | 3 | 81 |
| B41F | 4 | 91 | C06B | 4 | 90 | E01B | 4 | 92 | A62C | 3 | 84 | C10L | 3 | 81 |
| B61G | 4 | 91 | C06C | 4 | 90 | E01C | 4 | 92 | B04C | 3 | 84 | C12N | 3 | 81 |
| B62M | 4 | 91 | C06D | 4 | 90 | E04G | 4 | 92 | B27D | 3 | 84 | E02D | 3 | 81 |
| C06B | 4 | 91 | C09G | 4 | 90 | F03C | 4 | 92 | B31B | 3 | 84 | E06C | 3 | 81 |
| C06C | 4 | 91 | C12R | 4 | 90 | F15C | 4 | 92 | B31F | 3 | 84 | F03B | 3 | 81 |
| C06D | 4 | 91 | D02G | 4 | 90 | F17D | 4 | 92 | B41N | 3 | 84 | F15C | 3 | 81 |
| C09G | 4 | 91 | D21B | 4 | 90 | F41H | 4 | 92 | B44F | 3 | 84 | F23G | 3 | 81 |
| C12R | 4 | 91 | E01C | 4 | 90 | G03D | 4 | 92 | B61B | 3 | 84 | F23J | 3 | 81 |
| D21G | 4 | 91 | E04G | 4 | 90 | G04F | 4 | 92 | B61G | 3 | 84 | G04F | 3 | 81 |
| E01C | 4 | 91 | F17D | 4 | 90 | A43C | 3 | 93 | B66B | 3 | 84 | G04G | 3 | 81 |
| E04G | 4 | 91 | F26B | 4 | 90 | A62B | 3 | 93 | B67C | 3 | 84 | G07G | 3 | 81 |
| F17D | 4 | 91 | G03D | 4 | 90 | A62C | 3 | 93 | C06D | 3 | 84 | G09C | 3 | 81 |
| F24C | 4 | 91 | A23G | 3 | 91 | A63F | 3 | 93 | C10G | 3 | 84 | H03C | 3 | 81 |
| G03D | 4 | 91 | A43C | 3 | 91 | B27D | 3 | 93 | C10K | 3 | 84 | H04K | 3 | 81 |
| G06M | 4 | 91 | A62B | 3 | 91 | B31B | 3 | 93 | D01D | 3 | 84 | A01M | 2 | 82 |
| A62B | 3 | 92 | A62C | 3 | 91 | B31F | 3 | 93 | D21G | 3 | 84 | A41F | 2 | 82 |
| B21F | 3 | 92 | B21F | 3 | 91 | B61G | 3 | 93 | E02D | 3 | 84 | A46B | 2 | 82 |
| B23H | 3 | 92 | B23F | 3 | 91 | B61H | 3 | 93 | E06C | 3 | 84 | A47F | 2 | 82 |
| B25G | 3 | 92 | B25G | 3 | 91 | B62B | 3 | 93 | F03B | 3 | 84 | A62B | 2 | 82 |
| B31B | 3 | 92 | B27D | 3 | 91 | B66B | 3 | 93 | F03G | 3 | 84 | A63C | 2 | 82 |
| B31F | 3 | 92 | B31B | 3 | 91 | B67C | 3 | 93 | F24C | 3 | 84 | B02C | 2 | 82 |
| B44D | 3 | 92 | B31F | 3 | 91 | C06D | 3 | 93 | F26B | 3 | 84 | B42F | 2 | 82 |
| B61H | 3 | 92 | B44D | 3 | 91 | C10G | 3 | 93 | G01Q | 3 | 84 | B62H | 2 | 82 |
| C05F | 3 | 92 | B61H | 3 | 91 | C10K | 3 | 93 | G04D | 3 | 84 | B62L | 2 | 82 |
| C10K | 3 | 92 | B62B | 3 | 91 | D01D | 3 | 93 | G07G | 3 | 84 | C01F | 2 | 82 |
| C11D | 3 | 92 | B68G | 3 | 91 | D05B | 3 | 93 | H04K | 3 | 84 | C05F | 2 | 82 |
| C14B | 3 | 92 | C05F | 3 | 91 | D06J | 3 | 93 | A01G | 2 | 85 | C08C | 2 | 82 |
| D04C | 3 | 92 | C10K | 3 | 91 | E04C | 3 | 93 | A41F | 2 | 85 | C09G | 2 | 82 |
| D06J | 3 | 92 | D06J | 3 | 91 | F03G | 3 | 93 | A46B | 2 | 85 | C10G | 2 | 82 |
| E02B | 3 | 92 | E01B | 3 | 91 | F24C | 3 | 93 | A63C | 2 | 85 | D04C | 2 | 82 |
| F03C | 3 | 92 | E04C | 3 | 91 | F26B | 3 | 93 | A63G | 2 | 85 | E02B | 2 | 82 |
| F26B | 3 | 92 | G02C | 3 | 91 | G04D | 3 | 93 | B02C | 2 | 85 | E03B | 2 | 82 |
| G02C | 3 | 92 | G04D | 3 | 91 | G07G | 3 | 93 | B42F | 2 | 85 | E04D | 2 | 82 |

| | | | | | | | | | | | | | | |
|---|---|---|---|---|---|---|---|---|---|---|---|---|---|---|
| G04C | 3 | 92 | A21C | 2 | 92 | A41F | 2 | 94 | B44B | 2 | 85 | F22D | 2 | 82 |
| G04D | 3 | 92 | A23L | 2 | 92 | A46B | 2 | 94 | B62H | 2 | 85 | F26B | 2 | 82 |
| G07D | 3 | 92 | A63C | 2 | 92 | A63C | 2 | 94 | B62L | 2 | 85 | F28B | 2 | 82 |
| A21C | 2 | 93 | A63G | 2 | 92 | A63G | 2 | 94 | C01F | 2 | 85 | F42D | 2 | 82 |
| A23L | 2 | 93 | B27C | 2 | 92 | B42F | 2 | 94 | C05F | 2 | 85 | G03G | 2 | 82 |
| A63C | 2 | 93 | B28C | 2 | 92 | B44B | 2 | 94 | C09G | 2 | 85 | G06D | 2 | 82 |
| A63G | 2 | 93 | B31D | 2 | 92 | B62L | 2 | 94 | D05C | 2 | 85 | A01N | 1 | 83 |
| B28C | 2 | 93 | B42F | 2 | 92 | B62M | 2 | 94 | E01H | 2 | 85 | A22B | 1 | 83 |
| B31D | 2 | 93 | B44B | 2 | 92 | C05F | 2 | 94 | E02B | 2 | 85 | A22C | 1 | 83 |
| B42F | 2 | 93 | D01D | 2 | 92 | C21C | 2 | 94 | E04D | 2 | 85 | A23G | 1 | 83 |
| B44B | 2 | 93 | D05B | 2 | 92 | D07B | 2 | 94 | F25C | 2 | 85 | A61D | 1 | 83 |
| B68G | 2 | 93 | D21H | 2 | 92 | D21H | 2 | 94 | F28B | 2 | 85 | A61P | 1 | 83 |
| C25C | 2 | 93 | E01H | 2 | 92 | E01H | 2 | 94 | F42D | 2 | 85 | A63G | 1 | 83 |
| D01D | 2 | 93 | E02B | 2 | 92 | E02B | 2 | 94 | G03G | 2 | 85 | B31B | 1 | 83 |
| D21H | 2 | 93 | F03C | 2 | 92 | E04D | 2 | 94 | G06D | 2 | 85 | B43K | 1 | 83 |
| E01H | 2 | 93 | F24C | 2 | 92 | F25C | 2 | 94 | A01M | 1 | 86 | B44D | 1 | 83 |
| F16S | 2 | 93 | F25C | 2 | 92 | F28B | 2 | 94 | A01N | 1 | 86 | B44F | 1 | 83 |
| F22D | 2 | 93 | F28B | 2 | 92 | G06J | 2 | 94 | A23G | 1 | 86 | B61H | 1 | 83 |
| F25C | 2 | 93 | F42D | 2 | 92 | G10H | 2 | 94 | A61D | 1 | 86 | B63C | 1 | 83 |
| F28B | 2 | 93 | G06J | 2 | 92 | G21D | 2 | 94 | B07B | 1 | 86 | B64G | 1 | 83 |
| F42C | 2 | 93 | G10H | 2 | 92 | A01G | 1 | 95 | B27N | 1 | 86 | B65F | 1 | 83 |
| F42D | 2 | 93 | G21D | 2 | 92 | A01M | 1 | 95 | B43K | 1 | 86 | B67B | 1 | 83 |
| G06J | 2 | 93 | A01G | 1 | 93 | A01N | 1 | 95 | B61H | 1 | 86 | B67C | 1 | 83 |
| G10H | 2 | 93 | A01N | 1 | 93 | A47H | 1 | 95 | B63C | 1 | 86 | C07K | 1 | 83 |
| G21D | 2 | 93 | A43D | 1 | 93 | A47J | 1 | 95 | B64G | 1 | 86 | C10N | 1 | 83 |
| A01G | 1 | 94 | A45F | 1 | 93 | A61D | 1 | 95 | B65F | 1 | 86 | C23D | 1 | 83 |
| A43D | 1 | 94 | A47H | 1 | 93 | B31D | 1 | 95 | C07K | 1 | 86 | D01D | 1 | 83 |
| A45F | 1 | 94 | A47J | 1 | 93 | B43K | 1 | 95 | C08C | 1 | 86 | D05B | 1 | 83 |
| A47D | 1 | 94 | A47K | 1 | 93 | B64G | 1 | 95 | C10M | 1 | 86 | D05C | 1 | 83 |
| A47H | 1 | 94 | A61D | 1 | 93 | B65C | 1 | 95 | D05B | 1 | 86 | D06B | 1 | 83 |
| A47J | 1 | 94 | A63F | 1 | 93 | B65F | 1 | 95 | D06B | 1 | 86 | D06P | 1 | 83 |
| A47K | 1 | 94 | B62L | 1 | 93 | C07K | 1 | 95 | D06P | 1 | 86 | E21C | 1 | 83 |
| A61D | 1 | 94 | B62M | 1 | 93 | D06P | 1 | 95 | E21C | 1 | 86 | F16T | 1 | 83 |
| A62C | 1 | 94 | B64G | 1 | 93 | D21G | 1 | 95 | F16T | 1 | 86 | F21L | 1 | 83 |
| A63F | 1 | 94 | B65C | 1 | 93 | F16T | 1 | 95 | F42C | 1 | 86 | G06M | 1 | 83 |
| B27C | 1 | 94 | B65F | 1 | 93 | F41A | 1 | 95 | G06M | 1 | 86 | G10H | 1 | 83 |
| B27D | 1 | 94 | B66D | 1 | 93 | F42C | 1 | 95 | G10H | 1 | 86 | G21C | 1 | 83 |
| B62L | 1 | 94 | C07K | 1 | 93 | F42D | 1 | 95 | G21C | 1 | 86 | A42B | 0 | 84 |
| B65C | 1 | 94 | C10N | 1 | 93 | A42B | 0 | 96 | A22B | 0 | 87 | A47K | 0 | 84 |
| B65F | 1 | 94 | D06P | 1 | 93 | A44C | 0 | 96 | A42B | 0 | 87 | A63H | 0 | 84 |
| B66D | 1 | 94 | D07B | 1 | 93 | A45F | 0 | 96 | A45F | 0 | 87 | B27N | 0 | 84 |
| C07K | 1 | 94 | D21G | 1 | 93 | F28G | 0 | 96 | A63H | 0 | 87 | B60V | 0 | 84 |
| C10N | 1 | 94 | E05G | 1 | 93 | | | | B60V | 0 | 87 | B66D | 0 | 84 |
| D06P | 1 | 94 | F16T | 1 | 93 | | | | B65C | 0 | 87 | D01H | 0 | 84 |
| D21F | 1 | 94 | F22G | 1 | 93 | | | | B66D | 0 | 87 | E01C | 0 | 84 |
| E01B | 1 | 94 | F41A | 1 | 93 | | | | D01H | 0 | 87 | F22G | 0 | 84 |
| E01D | 1 | 94 | F42C | 1 | 93 | | | | D04C | 0 | 87 | | | |

| | | | | | |
|------|---|----|------|---|----|
| E03B | 1 | 94 | G04F | 1 | 93 |
| E03F | 1 | 94 | A46B | 0 | 94 |
| E05G | 1 | 94 | B60F | 0 | 94 |
| F22G | 1 | 94 | C10J | 0 | 94 |
| F41A | 1 | 94 | C21C | 0 | 94 |
| A01C | 0 | 95 | E03F | 0 | 94 |
| A46B | 0 | 95 | E04D | 0 | 94 |
| B60F | 0 | 95 | F28G | 0 | 94 |
| C10J | 0 | 95 | | | |
| C21C | 0 | 95 | | | |
| D21C | 0 | 95 | | | |
| E04D | 0 | 95 | | | |
| F28G | 0 | 95 | | | |
| G21C | 0 | 95 | | | |

Source: own calculations.

**Table A. 2:** Sample firms

| No. | Firm | No. | Firm |
|-----|------|-----|------|
| 1 | ACTech GmbH | 78 | Lear Corporation Electrical and Electronics GmbH & Co. KG |
| 2 | Adam Opel AG | 79 | LEONI AG |
| 3 | ADC Automotive Distance Control Systems GmbH | 80 | Leoni Kabel Holding GmbH & Co. KG |
| 4 | AKsys GmbH | 81 | Lucas Varity GmbH |
| 5 | Alutec Metallwaren GmbH & Co. KG | 82 | MAHLE GmbH |
| 6 | AMI Doduco GmbH | 83 | MAHLE International GmbH |
| 7 | Astyx GmbH | 84 | Mann + Hummel GmbH |
| 8 | ATMEL Automotive GmbH | 85 | Marquardt GmbH |
| 9 | AUDI Aktiengesellschaft | 86 | Menzolit-Fibron GmbH |
| 10 | AVL Deutschland GmbH | 87 | Meteor Gummiwerke K.H. Bädje GmbH & Co. KG |
| 11 | Bayerische Motoren Werke Aktiengesellschaft | 88 | Metzeler Schaum GmbH |
| 12 | BBS International GmbH | 89 | Micronas GmbH |
| 13 | Behr GmbH & Co. KG | 90 | Muhr und Bender KG |
| 14 | Behr-Hella Thermocontrol GmbH | 91 | Neosid Pemetzrieder GmbH & Co. KG |
| 15 | Benteler Automobiltechnik GmbH | 92 | NOVEM Car Interior Design GmbH |
| 16 | Wilhelm Böllhoff GmbH & Co. KG | 93 | odelo GmbH |
| 17 | Brose Fahrzeugteile GmbH & Co. KG | 94 | Oechsler Aktiengesellschaft |
| 18 | Car Trim GmbH | 95 | Optrex Europe GmbH |
| 19 | Freudenberg Gruppe | 96 | OSRAM GmbH |
| 20 | ZF Electronics GmbH | 97 | OSRAM Opto Semiconductors GmbH |
| 21 | Conti Temic microelectronic GmbH | 98 | paragon Aktiengesellschaft |
| 22 | Continental AG | 99 | PEIKER acustic GmbH & Co. KG |
| 23 | Continental Automotive GmbH | 100 | Philips Technologie GmbH |
| 24 | Continental Teves AG & Co. oHG | 101 | Pierburg GmbH |
| 25 | ContiTech Vibration Control GmbH | 102 | Pilkington Automotive Deutschland GmbH |
| 26 | Daimler AG | 103 | Plouquet Textiles Zittau GmbH |
| 27 | DEUTZ Aktiengesellschaft | 104 | Progress-Werk Oberkirch Aktiengesellschaft |
| 28 | Dr. Ing. h.c. F. Porsche Aktiengesellschaft | 105 | REHAU AG + Co |
| 29 | EDAG GmbH & Co. KGaA | 106 | REINZ-Dichtungs-GmbH |
| 30 | Entwicklungsgesellschaft für Akustik (EFA) mit beschränkter Haftung | 107 | Reum GmbH & Co. Betriebs KG |
| 31 | ELMOS Semiconductor AG | 108 | Robert Bosch GmbH |
| 32 | ElringKlinger AG | 109 | Robert Seuffer GmbH & Co. KG |
| 33 | EMITEC Gesellschaft für Emissionstechnologie mbH | 110 | Schaeffler Holding GmbH & Co. KG |
| 34 | EPCOS AG | 111 | Schunk Kohlenstofftechnik GmbH |
| 35 | ERAS Gesellschaft für Entwicklung und Realisation Adaptiver Systeme mbH | 112 | SEMIKRON International GmbH |
| 36 | Erhard & Söhne GmbH | 113 | Sensitec GmbH |
| 37 | ESG Elektroniksystem- und Logistik-Gesellschaft mit beschränkter Haftung | 114 | SFC Energy AG |
| 38 | Faurecia Abgastechnik GmbH | 115 | SGL Carbon GmbH |
| 39 | Faurecia Innenraum Systeme GmbH | 116 | Siemens Aktiengesellschaft |
| 40 | Federal-Mogul Burscheid GmbH | 117 | Siemens VDO Automotive AG |
| 41 | FEV Motorentechnik GmbH | 118 | Sitronic Gesellschaft für elektrotechnische Aus rüstung mbH. & Co. KG |
| 42 | Flabeg GmbH & Co. KG | 119 | Stankiewicz Gesellschaft mit beschränkter Haftung |
| 43 | Fludicon GmbH | 120 | Strähle + Hess GmbH |
| 44 | Ford-Werke GmbH | 121 | Texas Instruments Deutschland GmbH |
| 45 | Gardner Denver Thomas GmbH | 122 | ThyssenKrupp Fahrzeugguss GmbH / Thyssen Krupp Automotive AG |
| 46 | Georg Fischer Automobilguss GmbH | 123 | ThyssenKrupp Bilstein Suspension GmbH |
| 47 | GETRAG Getriebe- und Zahnradfabrik Hermann Hagenmeyer GmbH & Cie KG | 124 | ThyssenKrupp Drauz Nothelfer GmbH |

| 48 | GRAMMER AG | 125 | ThyssenKrupp Umformtechnik GmbH |
|----|-----------|-----|-------------------------------|
| 49 | Grohmann Engineering GmbH | 126 | TI Automotive |
| 50 | Harman Becker Automotive Systems (Becker Division) GmbH | 127 | Ticona GmbH |
| 51 | HARTING Automotive GmbH & Co. KG | 128 | TMD Friction |
| 52 | HBPO GmbH | 129 | TRW Airbag Systems GmbH |
| 53 | Hella KGaA Hueck & Co. | 130 | TRW Deutschland GmbH |
| 54 | Honda Research Institute Europe GmbH | 131 | TRW Automotive Safety Systems GmbH |
| 55 | Honeywell Bremsbelag GmbH | 132 | Tyco Electronics AMP GmbH |
| 56 | Huf Hülsbeck & Fürst GmbH & Co. KG | 133 | Umicore AG & Co. KG |
| 57 | Huf Tools GmbH Velbert | 134 | UST Umweltsensortechnik GmbH |
| 58 | Hydro Aluminium Deutschland GmbH | 135 | VERITAS AG |
| 59 | IAV GmbH Ingenieurgesellschaft Auto und Verkehr | 136 | Vibracoustic GmbH & Co. KG |
| 60 | IBEO Automobile Sensor GmbH | 137 | Volkswagen AG |
| 61 | IFA - Technologies GmbH | 138 | WABCO GmbH |
| 62 | I. G. Bauerhin GmbH | 139 | Walter Söhner GmbH & Co. KG |
| 63 | imk automotive GmbH | 140 | Webasto AG |
| 64 | Infineon Technologies AG | 141 | Westfalia Presstechnik GmbH & Co. KG |
| 65 | ISE Automotive GmbH | 142 | W.E.T. Automotive Systems Aktiengesellschaft |
| 66 | J. Eberspächer GmbH & Co. KG | 143 | Wilhelm Manz GmbH & Co. KG |
| 67 | Jacob Composite GmbH | 144 | WKW Erbslöh Automotive GmbH |
| 68 | Jenoptik Optical Systems GmbH | 145 | W. L. Gore & Associates GmbH |
| 69 | Jenoptik Polymer Systems GmbH | 146 | XCELLSIS AG |
| 70 | Johann Borgers GmbH & Co. KG | 147 | ZF Friedrichshafen AG |
| 71 | Johnson Controls Hybrid and Recycling GmbH | 148 | ZF Lemförder GmbH |
| 72 | Johnson Controls Headliner GmbH | 149 | ZF Lenksysteme GmbH |
| 73 | Karosseriewerke Dresden GmbH | 150 | ContiTech AG (incl. ContiTech Luftfedersysteme GmbH, ContiTech Profile GmbH, ContiTech Schlauch GmbH, ContiTech Vibration Control GmbH) |
| 74 | Kathrein-Werke KG | 151 | odelo LED GmbH |
| 75 | KEIPER GmbH & Co. KG | 152 | Polytec Automotive GmbH & Co. KG |
| 76 | KraussMaffei Technologies GmbH | 153 | Polytec Interior GmbH |
| 77 | Langendorf Textil GmbH & Co. KG | | |

Source: own illustration.

**Table A. 3**: Model 1 (correlation matrix).

| | Density | Transitivity | TECH_DISTANCE | ABSORPTIVE_CAPACITY | KB_MODULARITY |
|---|---|---|---|---|---|
| Density | | | | | |
| Transitivity | -0.121 | | | | |
| TECH_DISTANCE | 0.256 | 0.106 | | | |
| ABSORPTIVE_CAPACITY | -0.397 | -0.293 | 0.333 | | |
| KB_MODULARITY | -0.115 | -0.020 | 0.192 | 0.307 | |

Source: own calculations.

**Table A. 4:** Model 2 (correlation matrix).

| | Density | Transitivity | GEO_DISTANCE | TECH_DISTANCE | SIZE | ABSORPTIVE_CAPACITY | KB_MODULARITY | COOP_EXPERIENCE | IND_EXPERIENCE |
|---|---|---|---|---|---|---|---|---|---|
| Density | | | | | | | | | |
| Transitivity | -0.152 | | | | | | | | |
| GEO_DISTANCE | 0.213 | -0.118 | | | | | | | |
| TECH_DISTANCE | 0.270 | 0.126 | -0.020 | | | | | | |
| SIZE | -0.057 | -0.031 | -0.046 | -0.078 | | | | | |
| ABSORPTIVE_CAPACITY | -0.201 | -0.271 | 0.005 | 0.285 | 0.309 | | | | |
| KB_MODULARITY | -0.115 | -0.028 | -0.017 | 0.219 | 0.021 | 0.210 | | | |
| COOP_EXPERIENCE | -0.016 | -0.072 | -0.012 | -0.032 | -0.082 | -0.557 | -0.074 | | |
| IND_EXPERIENCE | 0.049 | 0.038 | 0.001 | -0.009 | 0.308 | -0.080 | -0.010 | -0.027 | |

Source: own calculations.

**Table A. 5:** Model 3 (correlation matrix).

| | Density | Transitivity | GEO_DISTANCE | TECH_DISTANCE | SIZE | ABSORPTIVE_CAPACITY | KB_MODULARITY | COOP_EXPERIENCE | IND_EXPERIENCE |
|---|---|---|---|---|---|---|---|---|---|
| Density | | | | | | | | | |
| Transitivity | -0.195 | | | | | | | | |
| GEO_DISTANCE | 0.159 | -0.099 | | | | | | | |
| TECH_DISTANCE | 0.220 | 0.132 | -0.007 | | | | | | |
| SIZE | -0.018 | 0.107 | -0.102 | -0.032 | | | | | |
| ABSORPTIVE_CAPACITY | -0.200 | -0.077 | -0.036 | 0.288 | 0.243 | | | | |
| KB_MODULARITY | -0.041 | -0.078 | -0.003 | 0.153 | 0.020 | 0.226 | | | |
| COOP_EXPERIENCE | 0.070 | -0.218 | 0.094 | -0.037 | -0.106 | -0.589 | -0.065 | | |
| IND_EXPERIENCE | 0.049 | 0.027 | -0.056 | -0.010 | 0.335 | -0.045 | -0.030 | -0.037 | |

Source: own calculations.

**Table A. 6:** Full model elements.

| | |
|---|---|
| 6 | Observations |
| 153 | Actors |
| 1 | Dependent network variable |
| 1 | Constant actor covariate |
| 4 | Exogenous changing actor covariates |
| 1 | Constant dyadic covariates |
| 1 | Exogenous changing dyadic covariates |
| 1 | File with times of composition change |

Source: own calculations.

**Table A. 7:** Composition changes for node set actors.

| Actor | 130 | Leaves network at time | 3 |
|---|---|---|---|
| Actor | 146 | Leaves network at time | 1 |
| Actor | 151 | Joins network at time | 2 |
| Actor | 152 | Joins network at time | 2 |
| Actor | 153 | Joins network at time | 3 |

Source: own calculations.

**Table A. 8:** Information about covariates (full model).

| | | Minimum | Maximum | Mean |
|---|---|---|---|---|
| SIZE | | 1.0 | 3.0 | 1.333 |
| GEO_DISTANCE | | 0 | 6.8 | 5.613 |
| TECH_DISTANCE | | | | 0.675 |
| ABSORPTIVE_CAPACITY | Period 1 | 0 | 8.8 | 3.020 |
| | Period 2 | 0 | 8.8 | 3.116 |
| | Period 3 | 0 | 8.7 | 3.167 |
| | Period 4 | 0 | 8.7 | 3.213 |
| | Period 5 | 0 | 8.7 | 3.138 |
| | Overall | | | 3.131 |
| KB_MODULARITY | Period 1 | 0 | 1 | 0.354 |
| | Period 2 | 0 | 1 | 0.370 |
| | Period 3 | 0 | 1 | 0.363 |
| | Period 4 | 0 | 1 | 0.344 |
| | Period 5 | 0 | 1 | 0.306 |

| | | | | |
|---|---|---|---|---|
| | Overall | | | 0.347 |
| COOP_EXPERIENCE | Period 1 | 0 | 96 | 4.327 |
| | Period 2 | 0 | 105 | 4.765 |
| | Period 3 | 0 | 113 | 5.176 |
| | Period 4 | 0 | 125 | 5.987 |
| | Period 5 | 0 | 129 | 6.712 |
| | Overall | | | 5.393 |
| IND_EXPERIENCE | Period 1 | 0 | 6.1 | 3.578 |
| | Period 2 | 0.7 | 6.1 | 3.647 |
| | Period 3 | 0 | 6.1 | 3.653 |
| | Period 4 | 0 | 6.1 | 3.685 |
| | Period 5 | 0.7 | 6.1 | 3.737 |
| | Overall | | | 3.660 |

Source: own calculations.

## B. Appendix: Figures

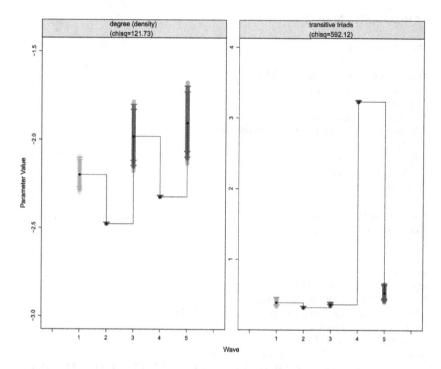

**Figure B. 1:** Test for time heterogeneity (small p values indicate time heterogeneity in the data) (Source: own calculations).

## C. Appendix: R-code

```
# *** LOGISTIC MAP ***

library(compiler)

logistic.map <- function(r, x, N, M){
# r: bifurcation parameter
# x: initial value
# N: number of iterations
# M: number of iteration points to be returned
z <- 1:N
z[1] <- x
  for(i in c(1:(N-1))){
    z[i+1] <- r *z[i] * (1 - z[i])
  }
# Return the last M iterations
z[c((N-M):N)]
}

logistic.map <- cmpfun(logistic.map)
my.r <- seq(2.7, 4, by=0.001)
N <- 2000; M <- 1000; start.x <- 0.1
orbit <- sapply(my.r, logistic.map, x=start.x, N=N, M=M)
Orbit <- as.vector(orbit)
r <- sort(rep(my.r, (M+1)))
plot(Orbit ~ r, pch=".", axes = F, col=rgb(0,0,0,0.05))
axis(side = 1, at = c(2.7, 2.8, 2.9, 3.0, 3.1, 3.2, 3.3, 3.4, 3.5, 3.7, 3.9, 4.0, 3.57, 3.83))
axis(side = 2, at = c(seq(0, 1, by=0.1)))


# *** MODULARITY INDICATOR ***

library (PCIT)

# Load patent database
patents = read.delim("patents.txt", header=TRUE)

# Define lists and start loop
ci.98 <- list()
ci.99 <- list()
ci.00 <- list()
ci.01 <- list()
ci.02 <- list()
ci <- list(ci.98, ci.99, ci.00, ci.01, ci.02)
y.start <- 1998
y.end <- 2002
```

```
b <- 1
while (y.end <= 2006)        {
                            # Select IPC sub-classes for relevant years and create
                            adjacency # matrices
                            x <- list()
                            i <- 1
                            while (i <= 153) {
                                                a <- subset (patents , ID == i &
                                                Prio_Year >= y.start & Prio_Year <=
                                                y.end, select = c(Appln_id,
                                                IPC.4.DIGIT))
                                                aT = as.matrix( table (droplevels(a)))
                                                aM = t(aT) %*% aT
                                                diag (aM) <- 0
                                                x [[i]] = assign (paste ("p", i, sep =""),
                                                aM)
                                                i <- i + 1
                                                }


                            # Replace tie strength values by "1" to calculate the
                            clustering #coefficients
                            j <- 1
                            while (j <= 153) {
                                                if (length(x[[j]]) > 0) {

                                                i <- 1
                                                while (i <= max(x[[j]])) {
                                                                        x[[j]][
                                                                        x[[j]]
                                                                        %in% i ]
                                                                        <- 1
                                                                        i <- i + 1
                                                                        }
                                                                        }
                                                j <- j + 1
                                                }

                            # Calculation of clustering coefficients
                            cc <- list ()
                            i <- 1
                            while (i <= 153) {
                                                c <- localClusteringCoefficient(x[[i]])
                                                cc [[i]] <- c
                                                i <- i + 1
                                                }
```

```
# Read patent data to calculate relative shares of IPC sub-
classes # in the patent portfolio
rp <- list()
i <- 1
while (i <= 153) {
                    a <- subset (patents , ID == i &
                    Prio_Year >= y.start & Prio_Year <=
                    y.end, select = c(Appln_id,
                    IPC.4.DIGIT))
                    h <- as.matrix (table (droplevels (a
                    [,2])))
                    sum (h)
                    rh <- h / sum (h)
                    rp [[i]] <- rh
                    i <- i + 1
                    }

# Calculation of the modularity indicators (mci)

i <- 1
while (i <= 153) {
                    cii <- cc [[i]] * rp [[i]]
                    ci [[b]] [i] <- sum (cii, na.rm = TRUE)

                    i <- i + 1
                    }
y.start <- y.start + 1
y.end <- y.end + 1
b <- b + 1
}

mci <- cbind (unlist(ci[[1]]), unlist(ci [[2]]), unlist(ci [[3]]), unlist(ci [[4]]), unlist(ci
[[5]]))




# *** STOCHASTIC ACTOR-BASED MODEL FOR NETWORK EVOLUTION ***


library(RSiena)
library (dplyr)

# Read in IDs of relevant actors
id <- read.delim("id.txt", header=TRUE)
```

```
# Construct networks from projects (bipartite data)
netzwerk_02 <- read.delim ("n02.txt", header = TRUE)
netzwerk_03 <- read.delim ("n03.txt", header = TRUE)
netzwerk_04 <- read.delim ("n04.txt", header = TRUE)
netzwerk_05 <- read.delim ("n05.txt", header = TRUE)
netzwerk_06 <- read.delim ("n06.txt", header = TRUE)
netzwerk_07 <- read.delim ("n07.txt", header = TRUE)

netzwerk_2002 <- netzwerk_02 %>%
   filter (Nummer.Akteur %in% as.numeric(id[,1]) ) %>%
   select (Nummer.Akteur, Projektnummer) %>%
   table () %>%
   as.matrix()

netzwerk_2003 <- netzwerk_03 %>%
   filter (Nummer.Akteur %in% as.numeric(id[,1]) ) %>%
   select (Nummer.Akteur, Projektnummer) %>%
   table () %>%
   as.matrix()

netzwerk_2004 <- netzwerk_04 %>%
   filter (Nummer.Akteur %in% as.numeric(id[,1]) ) %>%
   select (Nummer.Akteur, Projektnummer) %>%
   table () %>%
   as.matrix()

netzwerk_2005 <- netzwerk_05 %>%
   filter (Nummer.Akteur %in% as.numeric(id[,1]) ) %>%
   select (Nummer.Akteur, Projektnummer) %>%
   table () %>%
   as.matrix()

netzwerk_2006 <- netzwerk_06 %>%
   filter (Nummer.Akteur %in% as.numeric(id[,1]) ) %>%
   select (Nummer.Akteur, Projektnummer) %>%
   table () %>%
   as.matrix()

netzwerk_2007 <- netzwerk_07 %>%
   filter (Nummer.Akteur %in% as.numeric(id[,1]) ) %>%
   select (Nummer.Akteur, Projektnummer) %>%
   table () %>%
   as.matrix()

net_2002 <- netzwerk_2002 %*% t(netzwerk_2002)
```

```
net_2003 <- netzwerk_2003 %*% t(netzwerk_2003)
net_2004 <- netzwerk_2004 %*% t(netzwerk_2004)
net_2005 <- netzwerk_2005 %*% t(netzwerk_2005)
net_2006 <- netzwerk_2006 %*% t(netzwerk_2006)
net_2007 <- netzwerk_2007 %*% t(netzwerk_2007)

net_2002 <- ifelse (net_2002 > 0, 1, net_2002)
net_2003 <- ifelse (net_2003 > 0, 1, net_2003)
net_2004 <- ifelse (net_2004 > 0, 1, net_2004)
net_2005 <- ifelse (net_2005 > 0, 1, net_2005)
net_2006 <- ifelse (net_2006 > 0, 1, net_2006)
net_2007 <- ifelse (net_2007 > 0, 1, net_2007)

m <- matrix ( ,0, nrow = length(id [,1]), ncol = length(id [,1]))
rownames (m) <- id [,1]
colnames(m)<- id [,1]
m[is.na(m)] <- 0
match1 <- match(rownames(net_2002), rownames(m))
match2<- match(colnames(net_2002), colnames(m))
m2 <- m
m2[match1,match2] <- net_2002

match1 <- match(rownames(net_2003), rownames(m))
match2<- match(colnames(net_2003), colnames(m))
m3 <- m
m3[match1,match2] <- net_2003

match1 <- match(rownames(net_2004), rownames(m))
match2<- match(colnames(net_2004), colnames(m))
m4 <- m
m4[match1,match2] <- net_2004

match1 <- match(rownames(net_2005), rownames(m))
match2<- match(colnames(net_2005), colnames(m))
m5 <- m
m5[match1,match2] <- net_2005

match1 <- match(rownames(net_2006), rownames(m))
match2<- match(colnames(net_2006), colnames(m))
m6 <- m
m6[match1,match2] <- net_2006

match1 <- match(rownames(net_2007), rownames(m))
match2<- match(colnames(net_2007), colnames(m))
m7 <- m
m7[match1,match2] <- net_2007
```

```
# Read in adjacency matrices
madjazenz.1 <- m2
madjazenz.2 <- m3
madjazenz.3 <- m4
madjazenz.4 <- m5
madjazenz.5 <- m6
madjazenz.6 <- m7

# 1 Read in data for the absorptive capacity
absorp = read.delim("1_absortptive capacity.txt", header=TRUE)

# 2 Read in data for the technological distance
techdis.1 = read.delim("2_techdis2.txt", header=TRUE)
techdis.2 = read.delim("2_techdis3.txt", header=TRUE)
techdis.3 = read.delim("2_techdis4.txt", header=TRUE)
techdis.4 = read.delim("2_techdis5.txt", header=TRUE)
techdis.5 = read.delim("2_techdis6.txt", header=TRUE)

# 3 Read in data of the geographic distance
geo =  read.delim("3_geodis.txt", header=TRUE)

# 4 Read in data of the cooperation experience
experience = read.delim("4_experience.txt", header=TRUE)

# 5 Read in data of the size
size = read.delim("5_size.txt", header=TRUE)

# 6 Read in data of the industry experience
indexperience = read.delim("6_ind_experience.txt", header=TRUE)

# Transform data into matrices

# 1
mabsorp <- as.matrix(absorp)

# 2
mtechdis.1 <- as.matrix(techdis.1)
mtechdis.2 <- as.matrix(techdis.2)
mtechdis.3 <- as.matrix(techdis.3)
mtechdis.4 <- as.matrix(techdis.4)
mtechdis.5 <- as.matrix(techdis.5)

# 3
mgeo <- as.matrix(geo)
```

```
# 4
mexperience <- as.matrix(experience)

# 5
msize <- size

# 6
mindexperience <- as.matrix(indexperience)

# identify dependent variable of the model
cooperation <- sienaNet( array( c( madjazenz.1, madjazenz.2, madjazenz.3,
madjazenz.4, madjazenz.5, madjazenz.6 ), dim = c( 153, 153, 6 ) ) )

# identify independent variables (covariates) of the model

TECH_DISTANCE <- varDyadCovar(array( c( mtechdis.1, mtechdis.2, mtechdis.3,
mtechdis.4, mtechdis.5), dim = c( 153, 153, 5 ) ) )

GEO_DISTANCE <-  coDyadCovar (mgeo)

ABSORPTIVE_CAPACITY <- varCovar (mabsorp)

COOP_EXPERIENCE <- varCovar(mexperience)

IND_EXPERIENCE <- varCovar (mindexperience)

KB_MODULARITY <- varCovar (mci)

SIZE <- coCovar(msize[,1])


# Create composition change matrix
compositionchange <- sienaCompositionChangeFromFile("composition change.prn")


########## Model 0 (Base model) ##############

# Combine data for the analysis
mydata  <-  sienaDataCreate (cooperation)

# Create effects structure and include effects
myeff <- getEffects(mydata)
myeff <- includeEffects(myeff, density)
myeff <- includeEffects(myeff, transTriads)
```

```
# Estimation of parameters
mymodel.0  <- sienaModelCreate(projname = "final", modelType = 3, nsub=5,
n3=3000)
print01Report( mydata, myeff, modelname  = "descriptive" )
ans.0  <- siena07( mymodel.0, data = mydata, effects = myeff, returnDeps=TRUE)
```

########### Model 1 (includes knowledge related effects) ###############

```
# Combine data for the analysis
mydata <- sienaDataCreate (cooperation, TECH_DISTANCE,
ABSORPTIVE_CAPACITY, KB_MODULARITY, compositionchange)

# Create effects structure (inkl. time dummies for density and transitive triads)
myeff <- getEffects(mydata)

myeff <- includeEffects(myeff, density)
myeff <- includeEffects(myeff, transTriads)
myeff <- includeEffects(myeff, altX, interaction1 = "ABSORPTIVE_CAPACITY" )
myeff <- includeEffects(myeff, X, interaction1  = "TECH_DISTANCE" )
myeff <- includeEffects(myeff, altX, interaction1 = "KB_MODULARITY" )


# Estimation of parameters
mymodel.1 <- sienaModelCreate(projname = "final", modelType = 3, nsub=5,
n3=3000)
print01Report(mydata,  myeff, modelname  = "descriptive" )
ans.1  <-  siena07(mymodel.1, data = mydata, effects = myeff)
```

########### Model 2 (includes all effects) ###############

```
# Combine data for the analysis
mydata  <- sienaDataCreate (cooperation, TECH_DISTANCE,
ABSORPTIVE_CAPACITY,
KB_MODULARITY,GEO_DISTANCE,COOP_EXPERIENCE,IND_EXPERIENCE,
SIZE, compositionchange)

# Create effects structure
myeff <- getEffects(mydata)

myeff <- includeEffects(myeff, density)
myeff <- includeEffects(myeff, transTriads)
myeff  <- includeEffects(myeff, altX, interaction1 = "ABSORPTIVE_CAPACITY" )
myeff  <- includeEffects(myeff, X, interaction1 = "TECH_DISTANCE" )
```

```
myeff <- includeEffects(myeff, altX, interaction1 = "KB_MODULARITY" )
myeff <- includeEffects(myeff, X, interaction1 = "GEO_DISTANCE")
myeff <- includeEffects(myeff, altX, interaction1 = "COOP_EXPERIENCE" )
myeff <- includeEffects(myeff, altX, interaction1 = "IND_EXPERIENCE")
myeff <- includeEffects(myeff, altX, interaction1 = "SIZE" )


# Estimation of parameters
mymodel.2 <- sienaModelCreate(projname = "final", modelType = 3, nsub=5,
n3=3000)
ans.2 <- siena07( mymodel.2, data = mydata, effects = myeff, returnDeps=TRUE)

# Conduct a test for time heterogeneity and plot results
tt2  <- sienaTimeTest(ans.2)
summary (tt2)
plot(tt2, effects=1:9)

########### Model 3 (includes all effects and time dummies) ##############

# Combine data for the analysis
mydata  <- sienaDataCreate (cooperation, TECH_DISTANCE,
ABSORPTIVE_CAPACITY,
KB_MODULARITY,GEO_DISTANCE,COOP_EXPERIENCE,IND_EXPERIENCE,
SIZE, compositionchange)

# Create effects structure
myeff <- getEffects(mydata)

myeff <- includeEffects(myeff, density)
myeff <- includeEffects(myeff, transTriads)
myeff <- includeEffects(myeff, altX, interaction1 = "ABSORPTIVE_CAPACITY" )
myeff <- includeEffects(myeff, X, interaction1 = "TECH_DISTANCE" )
myeff <- includeEffects(myeff, altX, interaction1 = "KB_MODULARITY" )
myeff <- includeEffects(myeff, X, interaction1 = "GEO_DISTANCE")
myeff <- includeEffects(myeff, altX, interaction1 = "COOP_EXPERIENCE" )
myeff <- includeEffects(myeff, altX, interaction1 = "IND_EXPERIENCE")
myeff <- includeEffects(myeff, altX, interaction1 = "SIZE" )
myeff <- includeTimeDummy(myeff, density, timeDummy = "all")
myeff <- includeTimeDummy(myeff, transTriads, timeDummy = "all")


# Estimation of parameters
mymodel.3 <- sienaModelCreate(projname = "final", modelType = 3, nsub=5,
n3=3000)
ans.3 <- siena07( mymodel.3, data = mydata, effects = myeff, returnDeps=TRUE)
```

#### *** TEST FOR NORMAL DISTRIBUTION ***

```
# Histogram / Density-function of degree distribution and normal distribution overlay
h1 <- hist (deg1, breaks = 25, main = "Degree distribution of period 1998-2002", xlab
= "Degree", freq = FALSE)
# lines (density(deg1))
x <- seq(0, 200, 0.1)
curve(dnorm(x, mean = mean(deg1),sd = sd(deg1)), add = T)

# Shapiro-Test for normal distribution
shapiro.test (deg1)

 # QQ-Plots
library (qualityTools)

qqPlot(deg1, "normal", main = "Q-Q-Plot of period 1998-2002", xlab = "Quantiles for
degrees 1998-2002",ylab = "Quantiles from normal distribution", cex.lab=1.5 )


# *** ARC-DIAGRAM ***

library(arcdiagram)

adjazenz.1 = as.matrix(read.delim("matrix_98_02.txt", header=TRUE))
mode(adjazenz.1) <- "numeric" #bug
attributes.1 = as.matrix(read.delim("attributes_98_02.txt", header=FALSE))
p <- unique(sort(degree(graph_com.1),decreasing=TRUE))
(sort(degree(graph_com.1),decreasing=TRUE))
graph.1 <- delete.vertices(graph_com.1, V(graph_com.1)[ degree(graph_com.1) < p
[[10]]])

# get edgelist
edgelist = get.edgelist(graph.1)

# get edges value
weight = E(graph.1)$weight

# get vertex labels
vlabels = V(graph.1)$name

# get vertex sections
vsections = V(graph.1)$sections

 # get vertex fill color
vfill = V(graph.1)$fill
```

```
# get vertex border color
vborders = V(graph.1)$border

# get vertex degree
degrees = degree (graph_com.1, vlabels)

# data frame with vgroups, degree, vlabels and ind
x = data.frame(vsections, degrees, vlabels, ind=1:vcount(graph.1))

# arranging by vsections and degrees
y = arrange(x,desc(degrees), vlabels)

# get ordering 'ind'
new_ord = y$ind

# plot arc diagram
arcplot(edgelist, ordering=new_ord, labels=vlabels, cex.labels=0.8,
        show.nodes=TRUE, col.nodes=vfill, bg.nodes=vborders,
        cex.nodes = degrees/15, pch.nodes=21,
        lwd.nodes = 2, line=-0.5,
        col.arcs = hsv(0, 0, 0.2, 0.25), lwd.arcs = weight-50)


# *** MAP OF AUTOMOTIVE FIRMS ***

library (maptools)
library(ggmap)
library (sp)
library (rgdal)

read.csv("firm_map.csv",sep=";", header=T) -> firms
xx <- as.numeric(firms [,3])/1000000
yy <- as.numeric(firms [,4])/1000000
tt <- data.frame(cbind (xx , yy))
auto <- qmap ("Germany", zoom = 6, source = "stamen", maptype = "toner",
base_layer = ggplot (aes(x = yy, y = xx), data = tt)) + geom_point(size = 3, colour =
"red", alpha = 0.4)
auto
```

## Bibliography

Aghion, P. and Griffith, R. (2005). *Competition and growth: Reconciling theory and evidence. zeuthen lectures*, Cambridge: MIT Press.

Aghion, P. and Howitt, P. (1998). *Endogenous growth theory*, Cambridge: MIT Press.

Ahn, Y.-Y., Ahnert, S.E., Bagrow, J.P. and Barabási, A.-L. (2011). Flavor network and the principles of food pairing. *Scientific Reports*, 1, 196, DOI:10.1038/srep00196.

Ahuja, G. (2000a). Collaboration networks, structural holes, and innovation: A longitudinal study, *Administrative Science Quarterly*, **45** (3), pp. 425-455.

Ahuja, G. (2000b). The duality of collaboration: Inducements and opportunities in the formation of interfirm linkages, *Strategic Management Journal*, **21** (3), pp. 317-343.

Ahuja, G. and Katila, R. (2001). Technological acquisitions and the innovation performance of acquiring firms: A longitudinal study, *Strategic Management Journal*, **22** (3), pp. 197-220.

Almeida, P. and Kogut, B. (1999). Localization of knowledge and the mobility of engineers in regional networks, *Management science*, **45** (7), pp. 905-917.

Aristotle (1953). *Aristotle's metaphysics*, 2nd edn., Oxford: Clarendon Press.

Arndt, O. and Sternberg, R. (2000). Do manufacturing firms profit from intraregional innovation linkages? An empirical based answer, *European Planning Studies*, **8** (4), pp. 465-485.

Arora, A., Gambardella, A. and Rullani, E. (1997). Division of labour and the locus of inventive activity, *Journal of Management and Governance*, **1** (1), pp. 123-140.

Arrow, K.J. (1962). The economic implications of learning by doing, *The review of economic studies*, **29** (3), pp. 155-173.

Arrow, K.J. (1994). Methodological individualism and social knowledge, *The American Economic Review*, **84** (2), pp. 1-9.

Arthur, W.B. (1989). Competing technologies, increasing returns, and lock-in by historical events, *The economic journal*, **99** (394), pp. 116-131.

Arthur, W.B. (1999). Complexity and the economy, *Science*, **284** (5411), pp. 107-109.

Asheim, B.T. and Isaksen, A. (2002). Regional innovation systems: The integration of local 'sticky'and global 'ubiquitous' knowledge, *The Journal of Technology Transfer*, **27** (1), pp. 77-86.

Audretsch, D.B. and Feldman, M.P. (1996). Innovative clusters and the industry life cycle, *Review of Industrial Organization*, **11** (2), pp. 253-273.

Baker, W.E. (1990). Market networks and corporate behavior, *American journal of sociology*, **96** (3), pp. 589-625.

Bala, V. and Goyal, S. (2000). A noncooperative model of network formation, *Econometrica*, **68** (5), pp. 1181-1229.

Balconi, M., Breschi, S. and Lissoni, F. (2004). Networks of inventors and the role of academia: an exploration of Italian patent data, *Research Policy*, **33** (1), pp. 127-145.

Baldwin, C.Y. and Clark, K.B. (2000). *Design rules: The power of modularity*, Cambridge: MIT Press.

Baldwin, C. and Clark, K. (1997). Managing in an age of modularity. *Harvard business review*, **75** (5), pp. 84-93.

Balland, P.A. (2012). Proximity and the evolution of collaboration networks: Evidence from research and development projects within the global navigation satellite system (GNSS) industry, *Regional Studies*, **46** (6), pp. 741-756.

Balland, P.A., De Vaan, M. and Boschma, R. (2012). The dynamics of interfirm networks along the industry life cycle: The case of the global video game industry, 1987–2007, *Journal of Economic Geography*, **13** (5), pp. 1-25.

Barabási, A.L. (2005). Taming complexity, *Nature Physics*, **1** (2), pp. 68-70.

Barabási, A.L. (2007). The architecture of complexity, *IEEE Control Systems Magazine*, **27** (4), pp. 33-42.

Barabási, A.L. and Albert, R. (1999). Emergence of scaling in random networks, *Science*, **286** (5439), pp. 509-512.

Barabási, A.L. and Albert, R. (2002). Statistical mechanics of complex networks, *Reviews of modern physics*, **74** (1), pp. 47-97.

Barabási, A.L. and Oltvai, Z.N. (2004). Network biology: understanding the cell's functional organization, *Nature Reviews Genetics*, **5** (2), pp. 101-113.

Barnard, C. and Simon, H.A. (1976). *Administrative behavior. A study of decision-making processes in administrative organization*, New York: Free Press.

Barney, J. (1991). Firm resources and sustained competitive advantage, *Journal of management,* **17** (1), pp. 99-120.

Barro, R.J. and Sala-i-Martin, X. (2004). *Economic growth,* Cambridge: MIT Press.

Bassanini, A. and Scarpetta, S. (2001). The driving forces of economic growth: Panel data evidence for the OECD countries, *OECD Economic Studies,* **33** (2), pp. 9-56.

Bathelt, H. and Glückler, J. (2005). Resources in economic geography: From substantive concepts towards a relational perspective, *Environment and Planning,* **37** (9), pp. 1545-1563.

Baum, J.A.C., Shipilov, A.V. and Rowley, T.J. (2003). Where do small worlds come from? *Industrial and Corporate Change,* **12** (4), pp. 697-725.

Bell, M. and Albu, M. (1999). Knowledge systems and technological dynamism in industrial clusters in developing countries, *World Development,* **27** (9), pp. 1715-1734.

Benhabib, J., Spiegel, M. M. (1994). The role of human capital in economic development. Evidence from aggregate cross-country data, *Journal of Monetary Economics,* **34**, pp. 143-173.

Benner, M. and Waldfogel, J. (2008). Close to you? Bias and precision in patent-based measures of technological proximity, *Research Policy,* **37** (9), pp. 1556-1567.

Birchall, D.W., Tovstiga, G. and Chanaron, J. (2001). Capabilities in strategic knowledge sourcing and assimilation: a new look at innovation in the automotive industry, *International Journal of Automotive Technology and Management,* **1** (1), pp. 78-91.

BMBF (2008). Merkblatt für Antragsteller/Zuwendungsempfänger zur Zusammenarbeit der Partner von Verbundprojekten, *Bundesministerium für Bildung und Forschung, BMBF-Vordruck 0110/10.08.*

Borch, O.J. and Arthur, M.B. (1995). Strategic Networks Among Small Firms: Implications for Strategy Research Methodology, *Journal of Management Studies,* **32** (4), pp. 419-441.

Borgatti, S.P., Everett, M.G. and Freeman, L.C. (2002). Ucinet for Windows: Software for social network analysis.

Borgatti, S.P., Feld, S.L. (1994). How to test the strength of weak ties theory, *Connections,* **17**, 45-46.

Boschma, R. (2005). Proximity and innovation: A critical assessment, *Regional Studies,* **39** (1), pp. 61-74.

Boschma, R., Eriksson, R. and Lindgren, U. (2009). How does labour mobility affect the performance of plants? The importance of relatedness and geographical proximity, *Journal of Economic Geography*, **9** (2), pp. 169.

Boschma, R. and Iammarino, S. (2009). Related variety, trade linkages, and regional growth in Italy, *Economic Geography*, **85** (3), pp. 289-311.

Boschma, R.A. and ter Wal, A.L.J. (2007). Knowledge networks and innovative performance in an industrial district: The case of a footwear district in the South of Italy, *Industry & Innovation*, **14** (2), pp. 177-199.

Boschma, R.A. and Wenting, R. (2007). The spatial evolution of the British automobile industry: Does location matter? *Industrial and Corporate Change*, **16** (2), pp. 213-238.

Bottazzi, G. and Secchi, A. (2003). Why are distributions of firm growth rates tent-shaped? *Economics Letters*, **80** (3), pp. 415-420.

Bower, G.H. and Hilgard, E.R. (1981). *Theories of learning*, Englewood Cliffs: Prentice-Hall.

Brand, A. (1998). Knowledge management and innovation at 3M, *Journal of knowledge management*, **2** (1), pp. 17-22.

Brass, D.J., Galaskiewicz, J., Greve, H.R. and Tsai, W. (2004). Taking stock of networks and organizations: A multilevel perspective, *The Academy of Management Journal*, **47** (6), pp. 795-817.

Breschi, S. and Lissoni, F. (2001). Knowledge spillovers and local innovation systems: a critical survey, *Industrial and corporate change*, **10** (4), pp. 975-1005.

Breschi, S. and Lissoni, F. (2003). Mobility and social networks: Localised knowledge spillovers revisited, *CESPRI Working Paper No. 142*.

Breschi, S., Lissoni, F. and Malerba, F. (2003). Knowledge-relatedness in firm technological diversification, *Research Policy*, **32** (1), pp. 69-87.

Briggs, J. and Peat, F.D. (1990). *Die Entdeckung des Chaos: Eine Reise durch die Chaos-theorie*, München: Carl Hanser Verlag.

Broekel, T. and Graf, H. (2010). Structural properties of cooperation networks in Germany: From basic to applied research, *Jena Economic Research Papers, No. 2010, 78.*

Brusoni, S. and Prencipe, A. (2001). Unpacking the black box of modularity: Technologies, products and organizations, *Industrial and Corporate Change*, **10** (1), pp. 179-205.

Brynjolfsson, E. and Hitt, L.M. (2000). Beyond computation: Information technology, organizational transformation and business performance, *The Journal of Economic Perspectives*, **14** (4), pp. 23-48.

Buchanan, M. (2003). *Nexus: Small worlds and the groundbreaking theory of networks*, New York: WW Norton.

Buchmann, T., Hain, D., Kudic, M. and Müller, M. (2014). Exploring the co-evolutionary nature of innovation networks: New evidence from a stochastic actor-based approach, *IWH Discussion Papers, No. 2014, 1.*

Buchmann, T. and Pyka, A. (2012a). Innovation networks, in *Handbook on the economics and theory of the firm*, eds. M. Dietrich and J. Krafft, Cheltenham: Edward Elgar Publishing, pp. 466-484.

Buchmann, T. and Pyka, A. (2012b). Netzwerkindikatoren zur Bewertung der innovativen Leistungsfähigkeit, *Zeitschrift für Controlling und Innovationsmanagement*, **56** (S1), pp. 22-28.

Buchmann, T. and Pyka, A. (2015). The Evolution of Innovation Networks: The Case of a Publicly Funded German Automotive Network, *Economics of Innovation and New Technology*, **24** (1-2), pp. 114-139.

Burt, R.S. (1982). *Toward a structural theory of action: Network models of social structure, perception, and action*, New York: Academic Press.

Burt, R.S. (1995). *Structural holes: The social structure of competition*, Cambridge: Harvard University Press.

Burt, R.S. (2004). Structural Holes and Good Ideas, *American Journal of Sociology*, **110** (2), pp. 349-399.

Burt, R.S. and Knez, M. (1995). Kinds of third-party effects on trust, *Rationality and Society*, **7** (3), pp. 255-292.

Buskens, V. and Raub, W. (2002). Embedded trust: Control and learning, in *Group cohesion, trust and solidarity*, eds. E. Lawler and S. Thye, Bingley: Emerald Group Publishing Limited, pp. 167-202.

Camagni, R. (1991). Local "milieu", uncertainty and innovation networks: Towards a new dynamic theory of economic space, in *Innovation Networks: Spatial Perspectives*, ed. R. Camagni, London: Belhaven Press, pp. 121-142.

Cantner, U. and Pyka, A. (1998). Absorbing technological spillovers: simulations in an evolutionary framework, *Industrial and Corporate Change*, **7** (2), pp. 369-397.

Canton, E., Minne, B., Nieuwenhuis, A., Smid, B. and van der Steeg, M. (2005). Human capital, R&D, and competition in macroeconomic analysis, *CPB document No. 91.*

Cantwell, J. and Iammarino, S. (2003). Regional systems of innovation in Europe and the globalisation of technology, in *Multinational Corporations and European Regional Systems of Innovation*, eds. J. Cantwell and S. Iammarino, London: Routledge, pp. 7-21.

Chanaron, J. and Rennard, J. (2007). The automotive industry: a challenge to Schumpeter's innovation theory, in *Re-discovering Schumpeter: creative destruction evolving into "mode 3"*, eds. E. Carayannis and C. Ziemnowicz, New York: Sage, pp. 320-343.

Chen, P. (2008). Equilibrium illusion, economic complexity and evolutionary foundation in economic analysis, *Evolutionary and Institutional Economics Review*, **5** (1), pp. 81-127.

Chesbrough, H.W. (2003). The Era of Open Innovation, *MIT Sloan Management Review*, **44** (3), pp. 35-41.

Coase, R.H. (1937). The nature of the firm, *Economica*, **4** (16), pp. 386-405.

Cohen, M.D., Burkhart, R., Dosi, G., Egidi, M., Marengo, L., Warglien, M. and Winter, S. (1996). Routines and other recurring action patterns of organizations: Contemporary research issues, *Industrial and Corporate Change*, **5** (3), pp. 653-698.

Cohen, W.M. and Levinthal, D.A. (1989). Innovation and learning: The two faces of R&D, *The Economic Journal*, **99** (397), pp. 569-596.

Cohen, W.M. and Levinthal, D.A. (1990). Absorptive capacity: A new perspective on learning and innovation, *Administrative Science Quarterly*, **35** (1), pp. 128-152.

Cohen, W.M., Nelson, R.R. and Walsh, J.P. (2002). Links and impacts: The influence of public research on industrial R&D, *Management Science*, **48** (1), pp. 1-23.

Coleman, J.S. (1988). Social capital in the creation of human capital, *American journal of sociology*, **94** (S1), pp. 95-120.

Colombo, M.G. (2003). Alliance form: a test of the contractual and competence perspectives, *Strategic Management Journal*, **24** (12), pp. 1209-1229.

Costanza, R., Wainger, L., Folke, C. and Mäler, K.G. (1993). Modeling complex ecological economic systems, *Bioscience*, **43** (8), pp. 545-555.

Cowan, R. and Jonard, N. (2003). The dynamics of collective invention, *Journal of Economic Behavior and Organization*, **52** (4), pp. 513-532.

Cowan, R. and Jonard, N. (2004). Network structure and the diffusion of knowledge, *Journal of Economic Dynamics and Control*, **28** (8), pp. 1557-1575.

Cyert, R.M. and March, J.G. (1963). *A behavioral theory of the firm*, Englewood Cliffs: Prentice-Hall.

Czarnitzki, D., Doherr, T., Fier, A., Licht, G. and Rammer, C. (2003). Öffentliche Förderung der Forschungs- und Innovationsaktivitäten von Unternehmen in Deutschland. *Studien zum deutschen Innovationssystem Nr. 17–2003*, Zentrum für Europäische Wirtschafts-forschung (ZEW).

Czarnitzki, D., Ebersberger, B., and Fier, A. (2007). The relationship between R&D collaboration, subsidies, and R&D performance. *Journal of Applied Econometrics*, **22** (7), pp. 1347-1366.

Dahl, M.S. and Pedersen, C.Ø.R. (2004). Knowledge flows through informal contacts in industrial clusters: myth or reality? *Research Policy*, **33** (10), pp. 1673-1686.

Das, T.K. and Teng, B.S. (2000). A resource-based theory of strategic alliances, *Journal of Management*, **26** (1), pp. 31-61.

d'Aspremont, C. and Jacquemin, A. (1988). Cooperative and noncooperative R & D in duopoly with spillovers, *The American Economic Review*, **78** (5), pp. 1133-1137.

David, P.A. (1985). Clio and the Economics of QWERTY, *The American Economic Review*, **75** (2), pp. 332-337.

David, P.A. (2001). Path dependence, its critics and the quest for 'historical economics', in *Evolution and Path Dependence in Economic Ideas: Past and Present*, eds. P. Garrouste and S. Ioannides, Cheltenham: Edward Elgar Publishing, pp. 15-40.

Davis, G.F., Yoo, M. and Baker, W.E. (2003). The small world of the American corporate elite, 1982-2001, *Strategic Organization*, **1** (3), pp. 301-326.

Davis, J.A. (1970). Clustering and hierarchy in interpersonal relations: Testing two graph theoretical models on 742 sociomatrices, *American Sociological Review*, **35** (5), pp. 843-851.

De Rassenfosse, G. and van Pottelsberghe de la Potterie, B. (2009). A policy insight into the R&D-patent relationship, *Research Policy*, **38** (5), pp. 779-792.

Dean, L.G., Kendal, R.L., Schapiro, S.J., Thierry, B. and Laland, K.N. (2012). Identification of the social and cognitive processes underlying human cumulative culture, *Science*, **335** (6072), pp. 1114-1118.

Debackere, K., Luwel, M. and Veugelers, R. (1999). Can technology lead to a competitive advantage? A case study of Flanders using European patent data, *Scientometrics*, **44** (3), pp. 379-400.

Deloitte Consulting (2009). Konvergenz in der Automobilindustrie. Mit neuen Ideen Vorsprung sichern, *Industry Study*.

Deroian, F. (2002). Formation of social networks and diffusion of innovations, *Research policy*, **31** (5), pp. 835-846.

Dierickx, I. and Cool, K. (1989). Asset stock accumulation and sustainability of competitive advantage, *Management Science*, **35** (12), pp. 1504-1511.

Dilk, C., Gleich, R., Wald, A. and Motwani, J. (2008). Innovation networks in the automotive industry: An empirical study in Germany, *International Journal of Automotive Technology and Management*, **8** (3), pp. 317-330.

Dodgson, M. (1993). Learning, trust, and technological collaboration, *Human relations*, **46** (1), pp. 77-95.

Dollinger, M.J., Golden, P.A. and Saxton, T. (1997). The effect of reputation on the decision to joint venture, *Strategic Management Journal*, **18** (2), pp. 127-140.

Dopfer, K. (2005). *The evolutionary foundations of economics*, Cambridge: Cambridge University Press.

Dosi, G. (1988). Sources, procedures, and microeconomic effects of innovation, *Journal of economic literature*, **26** (3), pp. 1120-1171.

Dosi, G. and Nelson, R.R. (2010). Technical change and industrial dynamics as evolutionary processes, in *Handbook of the Economic of Innovation*, eds. B.H. Hall and N. Rosenberg, Amsterdam: North Holland, pp. 51-127.

Dosi, G. and Grazzi, M. (2010). On the nature of technologies: Knowledge, procedures, artifacts and production inputs, *Cambridge Journal of Economics*, **34** (1), pp. 173-184.

Doz, Y.L. (1996). The evolution of cooperation in strategic alliances: Initial conditions or learning processes?, *Strategic Management Journal*, **17** (S1), pp. 55-83.

Dyer, J.H. (1996). Specialized supplier networks as a source of competitive advantage: Evidence from the Auto industry, *Strategic Management Journal*, **17** (4), pp. 271-291.

Dyer, J.H. and Nobeoka, K. (2000). Creating and managing a high-performance knowledge-sharing network: The Toyota case, *Strategic Management Journal*, **21** (3), pp. 345-367.

Eisenhardt, K.M. and Schoonhoven, C.B. (1996). Resource-based view of strategic alliance formation: Strategic and social effects in entrepreneurial firms, *Organization Science,* **7** (2), pp. 136-150.

Eldredge, N. and Gould, S.J. (1972). Punctuated equilibria: An alternative to phyletic gradualism, in *Models in Paleobiology,* ed. T.J.M. Schopf, San Francisco: Freeman, Cooper and Company, pp. 82-115.

Engelsman, E. and Van Raan, A. (1991). Mapping of technology, A first exploration of knowledge diffusion amongst fields of technology, *Policy Studies on Technology and Economy, (BTE) Series, No. 15* .

Erdős, P. and Rényi, A. (1960). On the evolution of random graphs, *Publications of the Mathematical Institute of the Hungarian Academy of Sciences, No. 5,* pp. 17-61.

Estes, W.K. (1970). *Learning theory and mental development,* New York: Academic Press.

Ethiraj, S.K. and Levinthal, D. (2004). Modularity and innovation in complex systems, *Management Science,* **50** (2), pp. 159-173.

European Commission (2000). Presidency Conclusions, Lisbon European Council 23 and 24 March 2000, *Press Release.*

European Commission (2003). Building the knowledge society: Social and human capital interactions, *Commission Staff Working Document SEC (652).*

European Parliament (2012). Legislative resolution of 11 December 2012 on the proposal for a regulation of the European Parliament and of the Council implementing enhanced cooperation in the area of the creation of unitary patent protection.

Fagerberg, J. (1994). Technology and international differences in growth rates, *Journal of economic Literature,* **32** (3), pp. 1147-1175.

Feldman, M.P. (2000). Location and innovation: the new economic geography of innovation, spillovers, and agglomeration, in *The Oxford Handbook of Economic Geography,* eds. G. Clark L., M.P. Feldman and M.S. Gertler, Oxford: Oxford University Press, pp. 373-394.

Fleming, L. (2001). Recombinant uncertainty in technological search, *Management Science,* **47** (1), pp. 117-132.

Fleming, L., King, C. and Juda, A. (2007). Small worlds and innovation, *Organization Science,* **18** (6), pp. 938-954.

Fleming, L. and Sorenson, O. (2001). Technology as a complex adaptive system: Evidence from patent data, *Research Policy,* **30** (7), pp. 1019-1039.

Fornahl, D., Broekel, T. and Boschma, R. (2011). What drives patent performance of German biotech firms? The impact of R&D subsidies, knowledge networks and their location, *Papers in Regional Science,* **90** (2), pp. 395-418.

Freeman, C. (1991). Networks of innovators: A synthesis of research issues, *Research Policy,* **20** (5), pp. 499-514.

Freeman, C. (1994). The Economics of Technical Change, *Cambridge Journal of Economics,* **18** (5), pp. 463-514.

Freeman, L.C. (1979). Centrality in social networks conceptual clarification, *Social networks,* **1** (3), pp. 215-239.

Freeman, L.C., Romney, A.K. and Freeman, S.C. (1987). Cognitive structure and informant accuracy, *American anthropologist,* **89** (2), pp. 310-325.

Frenken, K., Van Oort, F. and Verburg, T. (2007). Related variety, unrelated variety and regional economic growth, *Regional Studies,* **41** (5), pp. 685-697.

Friedkin, N.E. and Johnsen, E.C. (1999). Social influence networks and opinion change, *Advances in Group Processes,* **16** (1), pp. 1-29.

Galaskiewicz, J. (1985). Interorganizational relations, *Annual Review of Sociology,* **11**, pp. 281-304.

Garcia-Pont, C. and Nohria, N. (2002). Local versus global mimetism: the dynamics of alliance formation in the automobile industry, *Strategic Management Journal,* **23** (4), pp. 307-321.

Gates, S. (1993). *Strategic alliances: Guidelines for successful management,* New York: Conference Board.

Gay, B. and Dousset, B. (2005). Innovation and network structural dynamics: Study of the alliance network of a major sector of the biotechnology industry, *Research policy,* **34** (10), pp. 1457-1475.

Georgesçu-Roegen, N. (1971). *The entropy law and the economic process,* Cambridge: Harvard University Press.

Ghoshal, S. (1987). Global strategy: An organizing framework, *Strategic Management Journal,* **8** (5), pp. 425-440.

Gilsing, V., Nooteboom, B., Vanhaverbeke, W., Duysters, G. and Van Den Oord, A. (2008). Network embeddedness and the exploration of novel technologies:

Technological distance, betweenness centrality and density, *Research Policy*, **37** (10), pp. 1717-1731.

Gintis, H. (2000). *Game theory evolving: A problem-centered introduction to modeling strategic behavior*, Princeton: Princeton University Press.

Giuliani, E. (2005). Cluster absorptive capacity: Why do some clusters forge ahead and others lag behind? *European Urban and Regional Studies*, **12** (3), pp. 269-288.

Giuliani, E. (2010). Network dynamics in regional clusters: The perspective of an emerging economy, *Papers in Evolutionary Economic Geography (PEEG), No. 1014*.

Giuliani, E. and Bell, M. (2005). The micro-determinants of meso-level learning and innovation: evidence from a Chilean wine cluster, *Research Policy*, **34** (1), pp. 47-68.

Glückler, J. (2007). Economic geography and the evolution of networks, *Journal of Economic Geography*, **7** (5), pp. 619-634.

Goerzen, A. and Beamish, P.W. (2005). The effect of alliance network diversity on multinational enterprise performance, *Strategic Management Journal*, **26** (4), pp. 333-354.

Gould, R.V. (2002). The origins of status hierarchies: A formal theory and empirical test, *American Journal of Sociology*, **107** (5), pp. 1143-1178.

Gould, S.J. and Vrba, E.S. (1982). Exaptation - a missing term in the science of form, *Paleobiology*, **8** (1), pp. 4-15.

Grabher, G. (1993). *The embedded firm. on the socioeconomics of industrial networks*, London: Routledge.

Granovetter, M.S. (1973). The strength of weak ties, *American journal of sociology*, **78** (6), pp. 1360-1380.

Granovetter, M.S. (1983). The strength of weak ties: A network theory revisited, *Sociological theory*, **1** (1), pp. 201-233.

Granovetter, M.S. (1985). Economic action and social structure: the problem of embeddedness, *American Journal of Sociology*, **91** (3), pp. 481-510.

Grant, R.M. (1991). The resource-based theory of competitive advantage: Implications for strategy formulation, *California Management Review*, **33**, pp. 114-135.

Grant, R.M. (1996). Toward a knowledge-based theory of the firm, *Strategic Management Journal,* **17** (Winter Special Issue), pp. 109-122.

Griliches, Z. (1990). Patent statistics as economic indicators: a survey, *Journal of economic literature,* **28** (4), pp. 1661-1707.

Griliches, Z. (1992). The search for R&D spillovers, *NBER Working Paper, No. 3768.*

Gulati, R. (1995a). Does familiarity breed trust? The implications of repeated ties for contractual choice in alliances, *Academy of management journal,* **38** (1), pp. 85-112.

Gulati, R. (1995b). Social structure and alliance formation patterns: A longitudinal analysis, *Administrative Science Quarterly,* **40** (4), pp. 619-652.

Gulati, R. (1998). Alliances and networks, *Strategic Management Journal,* **19** (4), pp. 293-317.

Gulati, R. and Gargiulo, M. (1999). Where Do Interorganizational Networks Come From? *American Journal of Sociology,* **104** (5), pp. 1439-1493.

Gulati, R., Nohria, N. and Zaheer, A. (2000). Strategic networks, *Strategic Management Journal,* **21** (3), pp. 203-215.

Hagedoorn, J. (1993). Understanding the rationale of strategic technology partnering: Interorganizational modes of cooperation and sectoral differences, *Strategic Management Journal,* **14** (5), pp. 371-385.

Hagedoorn, J. (1995). Strategic technology partnering during the 1980s: Trends, networks and corporate patterns in non-core technologies, *Research Policy,* **24** (2), pp. 207-231.

Hagedoorn, J. (2002). Inter-firm R&D partnerships: an overview of major trends and patterns since 1960, *Research policy,* **31** (4), pp. 477-492.

Hagedoorn, J. and Cloodt, M. (2003). Measuring innovative performance: Is there an advantage in using multiple indicators? *Research Policy,* **32** (8), pp. 1365-1379.

Hagedoorn, J. and Duysters, G. (2002). Learning in dynamic inter-firm networks-the efficacy of multiple contacts, *Organization Studies,* **23** (4), pp. 525-548.

Hall, B.H., Jaffe, A.B. and Trajtenberg, M. (2001). The NBER patent citation data file: Lessons, insights and methodological tools, *NBER Working Paper, No. 8498.*

Hall, B.H., Jaffe, A.B. and Trajtenberg, M. (2005). Market value and patent citations: A first look, *The Rand Journal of Economics,* **36** (1), pp. 16-38.

Hall, R. (1992). The strategic analysis of intangible resources, *Strategic Management Journal,* **13** (2), pp. 135-144.

Hamel, G. (1991). Competition for competence and interpartner learning within international strategic alliances, *Strategic Management Journal,* **12** (S1), pp. 83-103.

Hansen, M.T. (1999). The search-transfer problem: The role of weak ties in sharing knowledge across organization subunits, *Administrative Science Quarterly,* **44** (1), pp. 82-111.

Hanson, J.R. and Krackhardt, D. (1993). Informal networks: The company behind the chart, *Harvard business review,* **71** (4), pp. 104-111.

Hanusch, H. and Pyka, A. (eds) (2007a). *Elgar companion to neo-Schumpeterian economics,* Cheltenham: Edward Elgar Publishing.

Hanusch, H. and Pyka, A. (2007b). Manifesto for Comprehensive Neo-Schumpeterian Economics, *History of Economic Ideas,* **15** (1), pp. 23-42.

Hedström, P. (2005). *Dissecting the social: On the principles of analytical sociology,* Cambridge: Cambridge University Press.

Henderson, R. and Cockburn, I. (1996). Scale, scope, and spillovers: the determinants of research productivity in drug discovery, *The Rand journal of economics,* **27** (1), pp. 32-59.

Hennart, J.F. (1988). A transaction costs theory of equity joint ventures, *Strategic Management Journal,* **9** (4), pp. 361-374.

Hergert, M. and Morris, D. (1988). Trends in international collaborative agreements in *Cooperative strategies in international business,* eds. F. Contractor and P. Lorange, Lexington: Lexington Books, pp. 99-109.

Hidalgo, C.A., Klinger, B., Barabási, A.L. and Hausmann, R. (2007). The product space conditions the development of nations, *Science,* **317** (5837), pp. 482-487.

Hite, J.M. and Hesterly, W.S. (2001). The evolution of firm networks: From emergence to early growth of the firm, *Strategic Management Journal,* **22** (3), pp. 275-286.

Hite, J.M. (2008). The evolution of strategic dyadic network ties: Strategically navigating bounded agency within multidimensional and dynamic dyadic relationships, in *Network strategy: Advances in strategic management,* eds. J.A.C. Baum and T.J. Rowley, Bingley: JAI/Emerald Group, pp. 133-170.

Hoang, H. and Rothaermel, F.T. (2005). The effect of general and partner-specific alliance experience on joint R&D project performance, *The Academy of Management Journal*, **48** (2), pp. 332-345.

Hodgson, G.M. (ed.) (1995). *Economics and biology*, Aldershot: Edward Elgar Publishing.

Hodgson, G.M. (1997). *Economics and evolution: Bringing life back into economics*, Michigan: University of Michigan Press.

Hodgson, G.M. (1998). The approach of institutional economics, *Journal of Economic Literature*, **36** (1), pp. 166-192.

Hodgson, G.M., Samuels, W.J. and Tool, M.R. (eds) (1994). *The Elgar Companion to Institutional and Evolutionary Economics*, Aldershot: Edward Elgar Publishing.

Hoekman, J., Frenken, K. and Van Oort, F. (2009). The geography of collaborative knowledge production in Europe, *The Annals of Regional Science*, **43** (3), pp. 721-738.

Hofer, C.W. and Schendel, D. (1978). *Strategy formulation: Analytical concepts*, St. Paul: West Publishing Company.

Holland, J.H. (1995). *Hidden order: How adaptation builds complexity*, Reading: Addison-Wesley.

Holland, P.W. and Leinhardt, S. (1971). Transitivity in structural models of small groups, *Comparative Group Studies*, **2** (2), pp. 107-124.

Holland, P.W. and Leinhardt, S. (1977). A dynamic model for social networks, *The Journal of Mathematical Sociology*, **5** (1), pp. 5-20.

Hölldobler, B. and Wilson, E.O. (1990). *The ants*, Cambridge: Harvard University Press.

Homan, G. (1951). *The human group*, Abingdon: Routledge.

Hoover, K.D. (2010). Idealizing reduction: The microfoundations of macroeconomics, *Erkenntnis*, **73** (3), pp. 329-347.

Howells, J.R.L. (2002). Tacit knowledge, innovation and economic geography, *Urban Studies*, **39** (5-6), pp. 871-884.

Hubert, L.J. (1987). *Assignment methods in combinatorial data analysis*, New York: Dekker.

Ibarra, H. (1993). Network centrality, power, and innovation involvement: Determinants of technical and administrative roles, *Academy of Management Journal*, **36** (3), pp. 471-501.

Jacobs, J. (1970). *The economy of cities*, London: Random House.

Jaffe, A.B. (1986). Technological Opportunity and Spillovers of R&D: Evidence from Firms' Patents, Profits, and Market Value, *The American Economic Review*, **76** (5), pp. 984-1001.

Jaffe, A.B. (1989a). Characterizing the "technological position" of firms, with application to quantifying technological opportunity and research spillovers, *Research Policy*, **18** (2), pp. 87-97.

Jaffe, A.B. (1989b). Real effects of academic research, *The American Economic Review*, **79** (5), pp. 957-970.

Jaffe, A.B., Trajtenberg, M. and Henderson, R. (1993). Geographic localization of knowledge spillovers as evidenced by patent citations, *The Quarterly Journal of Economics*, **108** (3), pp. 577-598.

Jansen, D. (2006). *Einführung in die Netzwerkanalyse : Grundlagen, Methoden, Forschungsbeispiele*, Wiesbaden: Verlag für Sozialwissenschaften.

Johnson, J.L., Cullen, J.B., Sakano, T. and Takenouchi, H. (1996). Setting the stage for trust and strategic integration in Japanese-US cooperative alliances, *Journal of International Business Studies*, **27** (5), pp. 981-1004.

Jones, C.I. (1995). R&D-based Models of Economic Growth, *The Journal of Political Economy*, **103** (4), pp. 759-784.

Jones, C., Hesterly, W.S. and Borgatti, S.P. (1997). A General Theory of Network Governance: Exchange Conditions and Social Mechanisms, *The Academy of Management Review*, **22** (4), pp. 911-945.

Kale, P., Dyer, J.H. and Singh, H. (2002). Alliance capability, stock market response, and long-term alliance success: The role of the alliance function, *Strategic Management Journal*, **23** (8), pp. 747-767.

Karl, F. and Jäger, B. (2011). *Statistik deutscher Patentanmeldungen auf dem Gebiet der Elektromobilität im Zeitraum von 2000 bis 2009*, München: Spartas.

Kash, D.E. and Rycroft, R. (2002). Emerging patterns of complex technological innovation, *Technological forecasting and social change*, **69** (6), pp. 581-606.

Keynes, J.M. (1937). The general theory of employment, *The Quarterly Journal of Economics*, **51** (2), pp. 209-223.

Kitching, J. and Blackburn, R. (1999). Management training and networking in small and medium-sized enterprises in three European regions: implications for business support, *Environment and Planning: Government and Policy,* **17** (5), pp. 621-635.

Knight, F.H. (1921). *Risk, uncertainty and profit,* Boston: Hart, Schaffner & Marx.

Knoben, J. and Oerlemans, L.A.G. (2006). Proximity and inter-organizational collaboration: A literature review, *International Journal of Management Reviews,* **8** (2), pp. 71-89.

Knoke, D. and Yang, S. (2008). *Social network analysis,* Thousand Oaks: Sage Publications.

Kogut, B. (1988). Joint ventures: theoretical and empirical perspectives, *Strategic Management Journal,* **9** (4), pp. 319-332.

Kogut, B. (2000). The network as knowledge: generative rules and the emergence of structure, *Strategic Management Journal,* **21** (3), pp. 405-425.

Kogut, B. and Walker, G. (2001). The small world of Germany and the durability of national networks, *American Sociological Review,* **66** (3), pp. 317-335.

Kogut, B. and Zander, U. (1992). Knowledge of the firm, combinative capabilities, and the replication of technology, *Organization science,* **3** (3), pp. 383-397.

Kogut, B. and Zander, U. (1993). Knowledge of the firm and the evolutionary theory of the multinational corporation, *Journal of International Business Studies,* **24** (4), pp. 625-645.

Kotabe, M., Parente, R. and Murray, J. Y. (2007). Antecedents and outcomes of modular production in the Brazilian automobile industry: A grounded theory, *Journal of International Business Studies,* **38** (1), pp. 84-106.

Krackhardt, D. (1987). QAP partialling as a test of spuriousness, *Social Networks,* **9** (2), pp. 171-186.

Krackhardt, D. (1988). Predicting with networks: Nonparametric multiple regression analysis of dyadic data, *Social Networks,* **10** (4), pp. 359-381.

Lane, P.J. and Lubatkin, M. (1998). Relative absorptive capacity and interorganizational learning, *Strategic Management Journal,* **19** (5), pp. 461-477.

Laumann, E.O., Marsden, P.V. and Prensky, D. (1992). The boundary specification problem in network analysis, in *Research Methods in Social Network Analysis,* eds. L.C. Freeman, D.R. White and A.K. Romney, New Brunswick: Transaction Publishers, pp. 61-87.

Lazarsfeld, P.F. and Merton, R.K. (1954). Friendship as a social process: A substantive and methodological analysis, in *Freedom and Control in Modern Society*, ed. M. Berger, New York: Van Nostrand, pp. 18-66.

Lazega, E., Mounier, L., Snijders, T. and Tubaro, P. (2012). Norms, status and the dynamics of advice networks: A case study, *Social Networks*, **34** (3), pp. 323-332.

Lazer, D., Pentland, A., Adamic, L., Aral, S., Barabási, A.L., Brewer, D., Christakis, N., Contractor, N., Fowler, J., Gutmann, M., Jebara, T., King, G., Macy, M., Roy, D. and Van Alstyne, M. (2009). Computational social science, *Science*, **323** (6), pp. 721-723.

Leamer, E.E. and Storper, M. 2001, The economic geography of the internet age, *NBER Working Paper, No. 8450*.

Lechner, C. and Dowling, M. (1999). The evolution of industrial districts and regional networks: The case of the biotechnology region Munich/Martinsried, *Journal of Management and Governance*, **3** (4), pp. 309-338.

Leenders, R.T.A.J. (1995). Models for network dynamics: A Markovian framework, *The Journal of Mathematical Sociology*, **20** (1), pp. 1-21.

Levin, R.C., Cohen, W.M. and Mowery, D.C. (1985). R & D appropriability, opportunity, and market structure: New evidence on some Schumpeterian hypotheses, *The American Economic Review*, **75** (2), pp. 20-24.

Li, T.Y. and Yorke, J.A. (1975). Period three implies chaos, *American mathematical monthly*, **82** (10), pp. 985-992.

Lindsay, P.H. and Norman, D.A. (1977). *Human information processing*, Orlando: Academic Press.

Lippman, S.A. and Rumelt, R.P. (1982). Uncertain imitability: An analysis of interfirm differences in efficiency under competition, *The Bell Journal of Economics*, **13** (2), pp. 418-438.

Loasby, B.J. (2001). Time, knowledge and evolutionary dynamics: Why connections matter, *Journal of Evolutionary Economics*, **11** (4), pp. 393-412.

Lucas, R.E. (1988). On the mechanics of economic development, *Journal of Monetary Economics*, **22** (1), pp. 3-42.

Lucas, R.E. (1990). Why doesn't capital flow from rich to poor countries? *The American Economic Review*, **80** (2), pp. 92-96.

Madhok, A. (1997). Cost, value and foreign market entry mode: The transaction and the firm, *Strategic Management Journal*, **18** (1), pp. 39-61.

Makadok, R. (2001). Toward a synthesis of the resource-based and dynamic-capability views of rent creation, *Strategic Management Journal*, **22** (5), pp. 387-401.

Malerba, F. (2002). Sectoral systems of innovation and production, *Research policy*, **31** (2), pp. 247-264.

Mankiw, N.G. (2000). *Makroökonomik*, Stuttgart: Schäffer-Poeschel Verlag.

Mankiw, N.G., Romer, D. and Weil, D. N. (1992). A contribution to the empirics of economic growth, *The Quarterly Journal of Economics*, **107** (2), pp. 407-437.

Maraut, S., Dernis, H., Webb, C., Spiezia, V. and Guellec, D. (2008). The OECD REGPAT database: a presentation, *OECD Science, Technology and Industry Working Papers, No. 2008/02*.

March, J.G. (1991). Exploration and exploitation in organizational learning, *Organization Science*, **2** (1), pp. 71-87.

Marquis, C. (2003). The pressure of the past: Network imprinting in intercorporate communities, *Administrative Science Quarterly*, **48** (4), pp. 655-689.

Marsden, P.V. (1990). Network data and measurement, *Annual review of sociology*, **16** (1), pp. 435-463.

Marshall, A. (1920). *Principles of economics: An introductory volume*, 8th edn, London: Macmillan.

Marsili, M., Vega-Redondo, F. and Slanina, F. (2004). The rise and fall of a networked society: A formal model, *Proceedings of the National Academy of Sciences of the United States of America*, **101** (6), pp. 1439-1442.

May, R.M. (1976). Simple mathematical models with very complicated dynamics, *Nature*, **261** (5560), pp. 459-467.

McDonald, M.L. and Westphal, J.D. (2003). Getting by with the advice of their friends: CEOs' advice networks and firms' strategic responses to poor performance, *Administrative Science Quarterly*, **48** (1), pp. 1-32.

McPherson, M., Smith-Lovin, L. and Cook, J.M. (2001). Birds of a feather: Homophily in social networks, *Annual Review of Sociology*, **27** (2001), pp. 415-444.

Medda, G., Piga, C. and Siegel, D.S. (2006). Assessing the returns to collaborative research: Firm-level evidence from Italy, *Economics of Innovation and New technology*, **15** (1), pp. 37-50.

Menger, C. (1871). *Grundsätze der Volkswirthschaftslehre: Erster, allgemeiner Theil*, Wien: Wilhelm Braumüller.

Mill, J.S. (1865, 2008). *Auguste comte and positivism*, Rockville: Serenity Publishers.

Miller, D. and Shamsie, J. (1996). The resource-based view of the firm in two environments: The Hollywood film studios from 1936 to 1965, *Academy of management Journal*, **39** (3), pp. 519-543.

Mitchell, J.C. (1969). The concept and use of social networks, in *Social networks in urban situations: analyses of personal relationships in Central African towns*, ed. J.C. Mitchell, Manchester: The University Press, pp. 1-50.

Mokyr, J. (1990). *The lever of riches: Technological creativity and economic progress*, Oxford: Oxford University Press.

Moldoveanu, M.C., Baum, J.A.C. and Rowley, T.J. (2003). Information regimes, information strategies and the evolution of interfirm network topologies, *Research in Multi-Level Issues*, **2**, pp. 221-264.

Molina-Morales, F.X. and Martínez-Fernández, M.T. (2009). Too much love in the neighborhood can hurt: How an excess of intensity and trust in relationships may produce negative effects on firms, *Strategic Management Journal*, **30** (9), pp. 1013-1023.

Morrison, A. (2008). Gatekeepers of knowledge within industrial districts: who they are, how they interact, *Regional Studies*, **42** (6), pp. 817-835.

Mowery, D.C., Oxley, J.E. and Silverman, B.S. (1996). Strategic alliances and interfirm knowledge transfer, *Strategic Management Journal*, **17** (Special Issue: Knowledge and the Firm), pp. 77-91.

Müller, C. (2005). *Einführung in die kommerzielle Biotechnologie*, Stuttgart: Steinbeis Edition.

Müller, M., Buchmann, T. and Kudic, M. (2014). Micro Strategies and Macro Patterns in the Evolution of Innovation Networks – An Agent-Based Simulation Approach, in *Simulating knowledge dynamics in innovation networks*, eds. N. Gilbert, P. Ahrweiler and A. Pyka, Heidelberg: Springer, pp. 73-95.

Narin, F., Noma, E. and Perry, R. (1987). Patents as indicators of corporate technological strength, *Research policy*, **16** (2), pp. 143-155.

Nelson, R.R. and Sampat, B.N. (2001). Making sense of institutions as a factor shaping economic performance, *Journal of Economic Behavior and Organization*, **44** (1), pp. 31-54.

Nelson, R.R. and Teece, D.J. (2010). A discussion with Richard Nelson on the contributions of Alfred Chandler, *Industrial and Corporate Change*, **19** (2), pp. 351-361.

Nelson, R.R. and Winter, S.G. (1982). *An evolutionary theory of economic change*, Cambridge: Belknap Press.

Newman, M.E.J. (2001a). Scientific collaboration networks. II. Shortest paths, weighted networks, and centrality, *Physical Review E*, **64** (1), pp. 016132.

Newman, M.E.J. (2001b). The structure of scientific collaboration networks, *Proceedings of the National Academy of Sciences*, **98** (2), pp. 404-409.

Newman, M.E.J. (2003). The Structure and Function of Complex Networks, *SIAM Review*, **45** (2), pp. 167-256.

Newman, M.E.J., Barabasi, A.L. and Watts, D.J. (2006). *The structure and dynamics of networks*, Princeton: Princeton University Press.

Newman, M.E.J., Strogatz, S.H. and Watts, D.J. (2001). Random graphs with arbitrary degree distributions and their applications, *Physical Review E*, **64** (2), pp. 026118.

Niosi, J. (2005). *Canada's regional innovation system: The science-based industries*, Quebec: McGill-Queen's University Press.

Nooteboom, B. and Gilsing, V.A. 2004, Density and strength of ties in innovation networks: a competence and governance view, *ERIM Report Series, No. ERS-2004-005-ORG*.

OECD (1996). The Knowledge-based Economy, *General Distribution OCDE/GD*, **96** (102).

OECD (2003). Developments in growth literature and their relevance for simulation models, *Working Party on Global and Structural Policies*.

OECD (2005). *SME and entrepreneurship outlook*, Paris: OECD.

OECD (June 2010). REGPAT database.

Oliver Wyman (2012), *FAST 2025 – future automotive industry structure*, München.

Oliver, A.L. (2001). Strategic alliances and the learning life-cycle of biotechnology firms, *Organization Studies*, **22** (3), pp. 467-489.

Oliver, C. (1990). Determinants of interorganizational relationships: Integration and future directions, *The Academy of Management Review*, **15** (2), pp. 241-265.

Orsenigo, L., Pammolli, F. and Riccaboni, M. (2001). Technological change and network dynamics. Lessons from the pharmaceutical industry, *Research Policy*, **30** (3), pp. 485-508.

Owen-Smith, J. and Powell, W.W. (2004). Knowledge networks as channels and conduits: The effects of spillovers in the Boston biotechnology community, *Organization Science,* **15** (1), pp. 5-21.

Oxley, J.E. and Sampson, R.C. (2004). The scope and governance of international R&D alliances, *Strategic Management Journal,* **25** (8-9), pp. 723-749.

Parkhe, A. (1993). Strategic alliance structuring: A game theoretic and transaction cost examination of interfirm cooperation, *The Academy of Management Journal,* **36** (4), pp. 794-829.

Pavitt, K. (1984). Sectoral patterns of technical change: towards a taxonomy and a theory, *Research policy,* **13** (6), pp. 343-373.

Pavitt, K. (1985). Patent statistics as indicators of innovative activities: Possibilities and problems, *Scientometrics,* **7** (1), pp. 77-99.

Penrose, E.T. (1959). *The theory of the growth of the firm,* Oxford: Oxford University Press.

Perez, C. (2010). Technological revolutions and techno-economic paradigms, *Cambridge Journal of Economics,* **34** (1), pp. 185-202.

Peri, G. (2005). Determinants of knowledge flows and their effect on innovation, *Review of Economics and Statistics,* **87** (2), pp. 308-322.

Peteraf, M.A. (1993). The cornerstones of competitive advantage: A resource-based view, *Strategic Management Journal,* **14** (3), pp. 179-191.

Pfeffer, J. (1978). *The external control of organizations: A resource dependent perspective,* New York: Harper & Row.

Pisano, G.P. (1989). Using equity participation to support exchange: Evidence from the biotechnology industry, *Journal of Law, Economics, and Organization,* **5** (1), pp. 109-126.

Pittaway, L., Robertson, M., Munir, K., Denyer, D. and Neely, A. (2004). Networking and innovation: A systematic review of the evidence, *International Journal of Management Reviews,* **5** (3-4), pp. 137-168.

Podolny, J.M. (1993). A status-based model of market competition, *American journal of sociology,* **98** (4), pp. 829-872.

Podolny, J.M. (1994). Market uncertainty and the social character of economic exchange, *Administrative Science Quarterly,* **39** (3), pp. 458-483.

Poincaré, H. (1885). Sur l'équilibre d'une masse fluide animée d'un mouvement de rotation, *Acta mathematica,* **7** (1), pp. 259-380.

Polanyi, M. (1967). *The tacit dimension,* New York: Anchor Books.

Popielarz, P.A. and McPherson, J.M. (1995). On the edge or in between: Niche position, niche overlap, and the duration of voluntary association memberships, *American Journal of Sociology,* **101** (3), pp. 698-720.

Powell, W. W. (1990). Neither Market Nor Hierarchy: Network Forms of Organization, *Research in Organizational Behavior,* **12**, pp. 295-336.

Powell, W.W. and Brantley, P. (1992). Competitive cooperation in biotechnology: Learning through networks?, in *Networks and Organizations* , eds. N. Nohria and R.G. Eccles, Boston: Harvard Business School Press, pp. 366-394.

Powell, W.W., Koput, K.W. and Smith-Doerr, L. (1996). Interorganizational collaboration and the locus of innovation: Networks of learning in biotechnology. *Administrative Science Quarterly,* **41** (1), pp. 116-145.

Powell, W.W. and Smith-Doerr, L. (1994). Networks and economic life, in *The Handbook of Economic Sociology,* eds. N.J. Smelser and R. Swedberg, Princeton: Princeton University Press, pp. 368-402.

Powell, W.W., White, D.R., Koput, K.W. and Owen-Smith, J. (2005). Network Dynamics and Field Evolution: The Growth of Interorganizational Collaboration in the Life Sciences, *American journal of sociology,* **110** (4), pp. 1132-1205.

Price, D.S. (1976). A general theory of bibliometric and other cumulative advantage processes, *Journal of the American Society for Information Science,* **27** (5), pp. 292-306.

Prigogine, I. and Stengers, I. (1984). *Order out of chaos,* New York: Heinemann.

Pyka, A. (1997). Informal networking, *Technovation,* **17** (4), pp. 207-220.

Pyka, A. (1999). *Der kollektive Innovationsprozeß eine theoretische Analyse informeller Netzwerke und absorptiver Fähigkeiten,* Berlin: Duncker und Humblot.

Pyka, A. (2000). Informal Networks and Industry Life Cycles, *Technovation,* **20** (1), pp. 25-35.

Pyka, A. (2002). Innovation networks in economics: From the incentive-based to the knowledge-based approaches, *European Journal of Innovation Management,* **5** (3), pp. 152-163.

Pyka, A. and Fagiolo, G. (2005). Agent-based modelling: A methodology for neo-schumpeterian economics, in *The Elgar Companion to Neo-Schumpeterian Economics*, eds. H. Hanusch and A. Pyka, Cheltenham: Edward Elgar Publishing, pp. 467-487.

Pyka, A., Gilbert, N. and Ahrweiler, P. (2007). Simulating Knowledge-Generation and Distribution Processes in Innovation Collaborations and Networks, *Cybernetics and Systems: An International Journal*, **38** (7), pp. 667-693.

Pyka, A., Gilbert, N. and Ahrweiler, P. (2009). Agent-Based Modelling of Innovation Networks – The Fairytale of Spillovers, in *Innovation Networks: New Approaches in Modelling and Analyzing*, eds. A. Pyka and A. Scharnhorst, Berlin: Springer , pp. 101-126.

Pyka, A. and Hanusch, H. (eds) (2006). *Applied evolutionary economics and the knowledge-based economy*, Cheltenham: Edward Elgar Publishing.

Pyka, A. and Saviotti, P. (2005). The evolution of R&D networking in the biotech industries, *International Journal of Entrepreneurship and Innovation Management*, **5** (1), pp. 49-68.

Ramanathan, K., Seth, A. and Thomas, H. (1997). Explaining joint ventures: Alternative theoretical perspectives, in *Cooperative Strategies: Vol. 1. North American Perspectives*, eds. P.W. Beamish and J.P. Killing, San Francisco: New Lexington Press, pp. 51-85.

Raub, W. and Weesie, J. (1990). Reputation and efficiency in social interactions: An example of network effects, *American Journal of Sociology*, **96** (3), pp. 626-654.

Reagans, R. and McEvily, B. (2003). Network structure and knowledge transfer: The effects of cohesion and range, *Administrative Science Quarterly*, **48** (2), pp. 240-267.

Redlich, F. (1944). The leaders of the German steam-engine industry during the first hundred years, *The Journal of Economic History*, **4** (2), pp. 121-148.

Reed, R. and DeFillippi, R.J. (1990). Causal ambiguity, barriers to imitation, and sustainable competitive advantage, *Academy of management review*, **15** (1), pp. 88-102.

Rendell, L., Boyd, R., Cownden, D., Enquist, M., Eriksson, K., Feldman, M.W., Fogarty, L., Ghirlanda, S., Lillicrap, T. and Laland, K.N. (2010). Why Copy Others? Insights from the Social Learning Strategies Tournament, *Science*, **328** (5975), pp. 208-213.

Ricker, W.E. (1954). Stock and recruitment, *Journal of the Fisheries Board of Canada*, **11** (5), pp. 559-623.

Ring, P.S. and Van de Ven, A.H. (1992). Structuring cooperative relationships between organizations, *Strategic Management Journal,* **13** (7), pp. 483-498.

Ripley, R.M., Snijders, T.A. and Preciado, P. (2010). Manual for SIENA version 4.0, *University of Oxford, Department of Statistics & Nuffield College.*

Robbins, H. and Monro, S. (1951). A stochastic approximation method, *The Annals of Mathematical Statistics,* **22** (3), pp. 400-407.

Robins, G., Pattison, P., Kalish, Y. and Lusher, D. (2007). An introduction to exponential random graph (p) models for social networks, *Social Networks,* **29** (2), pp. 173-191.

Robinson, D.T. and Stuart, T.E. (2002). Just how incomplete are incomplete contracts? Evidence from biotech strategic alliances, *Working Paper, Columbia University.*

Rodrigues, M.J. (2003). *European policies for a knowledge economy,* Cheltenham: Edward Elgar Publishing.

Roland Berger (2010). *Interview,* 14.4.2010.

Romer, P.M. (1986). Increasing returns and long-run growth, *The Journal of Political Economy,* **94** (5), pp. 1002-1037.

Romer, P.M. (1990). Endogenous Technological Change, *Journal of Political Economy,* **95** (5), pp. 71-102.

Rosenberg, N. (1990). Why do firms do basic research (with their own money)? *Research policy,* **19** (2), pp. 165-174.

Rosenberg, N. (2009). Some critical episodes in the progress of medical innovation: An Anglo-American perspective, *Research Policy,* **38** (2), pp. 234-242.

Rost, K. (2011). The strength of strong ties in the creation of innovation, *Research policy,* **40** (4), pp. 588-604.

Rothaermel, F.T. and Boeker, W. (2008). Old technology meets new technology: Complementarities, similarities, and alliance formation, *Strategic Management Journal,* **29** (1), pp. 47-77.

Rowley, T., Behrens, D. and Krackhardt, D. (2000). Redundant governance structures: An analysis of structural and relational embeddedness in the steel and semiconductor industries, *Strategic Management Journal,* **21** (3), pp. 369-386.

Rumelt, R. (1997). Towards a Strategic Theory of the Firm, in *Resources, Firms and Strategies,* ed. N.J. Foss, Oxford: Oxford University Press, pp. 131-145.

Rycroft, R.W. and Kash, D.E. (2004). Self-organizing innovation networks: implications for globalization, *Technovation*, **24** (3), pp. 187-197.

Samuels, W.J. (1972). The scope of economics historically considered, *Land Economics*, **48** (3), pp. 248-268.

Sanchez, R. and Mahoney, J.T. (1996). Modularity, flexibility, and knowledge management in product and organization design, *Strategic Management Journal*, **17** (Winter Special Issue), pp. 63-76.

Saviotti, P.P. (2004). Considerations about the production and utilization of knowledge, *Journal of Institutional and Theoretical Economics*, **160** (1), pp. 100-121.

Saviotti, P. P. (2009). Knowledge networks: Structure and dynamics, in *Innovation Networks: Understanding Complex Systems*, eds. A. Pyka and A. Scharnhorst, Berlin: Springer, pp. 19-41.

Sayer, A. (1991). Behind the locality debate: Deconstructing geography's dualisms, *Environment and Planning A*, **23** (2), pp. 283-308.

Schilling, M.A. (2000). Toward a general modular systems theory and its application to interfirm product modularity, *Academy of Management Review*, **25** (2), pp. 312-334.

Schoen, A., Villard, L., Laurens, P., Cointet, J., Heimeriks, G. and Alkemade, F. (2012). The network structure of technological developments; Technological distance as a walk on the technology map, *Paper presented at the Science & Technology Indicators (STI) 2012 Montreal*.

Schön, B. and Pyka, A. (2012). A taxonomy of innovation networks, *FZID discussion papers, No. 42-2012*.

Schumpeter, J.A. (1911). *Theorie der wirtschaftlichen Entwicklung*, Berlin: Duncker & Humblot.

Schumpeter, J.A. (1939). *Business cycles. A theoretical, historical and statistical analysis of the capitalist process*, New York: McGraw-Hill Book Company.

Schumpeter, J.A. (1942). *Capitalism, socialism and democracy*, New York: Harper.

Scott, J. (2000). *Social network analysis: A handbook*, London: Sage Publications.

Shapiro, S.S. and Wilk, M.B. (1965). An analysis of variance test for normality (complete samples), *Biometrika*, **52** (3/4), pp. 591-611.

Shaw, B. (1993). Formal and informal networks in the UK medical equipment industry, *Technovation*, **13** (6), pp. 349-365.

Shionoya, Y. (2007). Schumpeter and Evolution: A Philosophical Interpretation, *History of Economic Ideas*, **15** (1), pp. 65-80.

Sianesi, B. and Reenen, J. (2003). The returns to education: Macroeconomics, *Journal of economic surveys*, **17** (2), pp. 157-200.

Simon, H.A. (1956). Rational choice and the structure of the environment, *Psychological Review*, **63** (2), pp. 129-138.

Simon, H.A. (1962). The architecture of complexity, *Proceedings of the American Philosophical Society*, **106** (6), pp. 467-482.

Simon, H.A. (1995). Near Decomposability and complexity: How a mind resides in a brain, in *The mind, the brain and complex adaptive systems. SFI studies in the sciences of complexity, Vol. XXII*, eds. H. Morowitz and J. Singer, Reading: Addison-Wesley, pp. 25-43.

Skvoretz, J. and Willer, D. (1991). Power in exchange networks: Setting and structural variations, *Social Psychology Quarterly*, **54** (3), pp. 224-238.

Smelser, N.J. and Swedberg, R. (1994). The sociological perspective on the economy, in *The handbook of economic sociology*, eds. N.J. Smelser and R. Swedberg, Princeton: Princeton University Press, pp. 3-26.

Snijders, T.A.B. (1996). Stochastic actor-oriented models for network change, *Journal of Mathematical Sociology*, **21** (1-2), pp. 149-172.

Snijders, T.A.B. (2001). The statistical evaluation of social network dynamics, *Sociological methodology*, **31** (1), pp. 361-395.

Snijders, T.A.B. (2005). Models for longitudinal network data, in *Models and Methods in Social Network Analysis*, eds. P.J. Carrington, J. Scott and S. Wasserman, Cambridge: Cambridge University Press, pp. 215-247.

Snijders, T.A.B. (2008). Statistical modeling of dynamics of non-directed networks, *Presentation at the XXV International Sunbelt Social Networks Conference, Redondo Beach (Los Angeles), February 16-20. 2005. Revised version.*

Snijders, T.A.B., Van de Bunt, G.G. and Steglich, C.E.G. (2010). Introduction to stochastic actor-based models for network dynamics, *Social Networks*, **32** (1), pp. 44-60.

Snijders, T.A.B., Lomi, A. and Torló, V.J. (2013). A model for the multiplex dynamics of two-mode and one-mode networks, with an application to employment preference, friendship, and advice, *Social networks*, **35** (2), pp. 265-276.

Solow, R.M. (1956). A contribution to the theory of economic growth, *The Quarterly Journal of Economics,* **70** (1), pp. 65-94.

Sonn, J.W. and Storper, M. (2008). The increasing importance of geographical proximity in knowledge production: an analysis of US patent citations, 1975-1997, *Environment and planning A,* **40** (5), pp. 1020-1039.

Sorenson, O., Rivkin, J.W. and Fleming, L. (2006). Complexity, networks and knowledge flow, *Research Policy,* **35** (7), pp. 994-1017.

Spencer, H. (1874). *The principles of biology,* London: Williams & Norgate.

Spencer, J.W. (2003). Firms' knowledge-sharing strategies in the global innovation system: empirical evidence from the flat panel display industry, *Strategic Management Journal,* **24** (3), pp. 217-233.

Spender, J.C. (1996). Making knowledge the basis of a dynamic theory of the firm, *Strategic Management Journal,* **17** (Special issue: Knowledge and the firm), pp. 45-62.

Stahlecker, T., Lay, G. and Zanker, C. (2010). *Elektromobilität: Zulieferer für den Strukturwandel gerüstet? Status quo und Handlungsempfehlungen für den Automobilstandort Metropolregion Stuttgart,* Studie im Auftrag der IHK Stuttgart.

Staiger, T.J., Gleich, R. and Dilk, C. (2006). Innovationsnetzwerke in der Automobilindustrie, *Zeitschrift für Controlling und Innovationsmanagement,* **1** (3), pp. 34-39.

Sterlacchini, A. and Venturini, F. (2006). Is Europe becoming a knowledge-driven economy? Evidence from EU developed regions, *Working Paper 253, Universita' Politecnica delle Marche (I), Dipartimento di Economia* .

Stevens, C. (1996). The knowledge-driven economy, *OECD Observer,* **200** (1), pp. 6-10.

Stokman, F.N. and Doreian, P. (1997). Evolution of social networks: Processes and principles, in *Evolution of Social Networks,* eds. P. Doreian and F.N. Stokman, New York: Routledge, pp. 233-250.

Storper, M. and Venables, A.J. (2004). Buzz: face-to-face contact and the urban economy, *Journal of economic geography,* **4** (4), pp. 351-370.

Sturgeon, T.J., Van Biesebroeck, J. and Gereffi, G. (2008). Value chains, networks and clusters: Reframing the global automotive industry, *Journal of Economic Geography,* **8** (3), pp. 297-321.

Suda, T., Itao, T. and Matsuo, M. (2010). The bio-networking architecture: The biologically inspired approach to the design of scalable, adaptive, and survivable/available network applications, *Science, 327*, pp. 439-441.

Swan, T.W. (1956). Economic growth and capital accumulation, *Economic Record, 32* (2), pp. 334-361.

Tabuchi, T. (1998). Urban agglomeration and dispersion: A synthesis of Alonso and Krugman. *Journal of Urban Economics, 44* (3), pp. 333-351.

Takeishi, A. (2001). Bridging inter-and intra-firm boundaries: Management of supplier involvement in automobile product development, *Strategic Management Journal, 22* (5), pp. 403-433.

Takeishi, A. (2002). Knowledge partitioning in the interfirm division of labor: The case of automotive product development, *Organization Science, 13* (3), pp. 321-338.

Teece, D.J. (1986). Profiting from technological innovation: Implications for integration, collaboration, licensing and public policy, *Research policy, 15* (6), pp. 285-305.

Teece, D.J. (1992). Competition, cooperation, and innovation: Organizational arrangements for regimes of rapid technological progress, *Journal of Economic Behavior & Organization, 18* (1), pp. 1-25.

Teece, D.J. and Pisano, G. (1994). The dynamic capabilities of firms: An introduction, *Industrial and Corporate Change, 3* (3), pp. 537-556.

Ter Wal, A.L.J. (2014). The spatial dynamics of the inventor network in German biotechnology. Geographical proximity versus triadic closure. *Journal of Economic Geography, 14* (3), pp. 589-620.

Trajtenberg, M. (1990). A penny for your quotes: Patent citations and the value of innovations, *The Rand journal of economics, 21* (1), pp. 172-187.

Tsai, W. (2001). Knowledge transfer in intraorganizational networks: Effects of network position and absorptive capacity on business unit innovation and performance, *Academy of Management Journal, 44* (5), pp. 996-1004.

Tsang, E.W.K. (2000). Transaction cost and resource-based explanations of joint ventures: A comparison and synthesis, *Organization Studies, 21* (1), pp. 215-242.

Udehn, L. (2002). The changing face of methodological individualism, *Annual Review of Sociology, 28*, pp. 479-508.

Utterback, J.M. (1995). Dominant designs and the survival of firms, *Strategic Management Journal, 16* (6), pp. 415-430.

Uzzi, B. (1997). Social structure and competition in interfirm networks: The paradox of embeddedness, *Administrative Science Quarterly*, **42** (1), pp. 35-67.

Uzzi, B. and Spiro, J. (2005). Collaboration and creativity: The small world problem, *American Journal of Sociology*, **111** (2), pp. 447-504.

Valente, T.W. (1996). Social network thresholds in the diffusion of innovations, *Social Networks*, **18** (1), pp. 69-89.

Van Biesebroeck, J. (2003). Productivity dynamics with technology choice: An application to automobile assembly, *The Review of Economic Studies*, **70** (1), pp. 167-198.

Van de Bunt, G.G. and Groenewegen, P. (2007). An actor-oriented dynamic network approach, *Organizational Research Methods*, **10** (3), pp. 463-482.

Van de Ven, A.H. (1976). On the nature, formation, and maintenance of relations among organizations, *The Academy of Management Review*, **1** (4), pp. 24-36.

Van Den Bergh, J.C.J.M. and Gowdy, J.M. (2003). The microfoundations of macroeconomics: An evolutionary perspective, *Cambridge Journal of Economics*, **27** (1), pp. 65-84.

Verhulst, P.F. (1845). Recherches mathématiques sur la loi d'accroissement de la population, *Mémoires de l'Académie Royale des Sciences et des Belles-Lettres de Bruxelles*, **18**, pp. 14-54.

Verhulst, P.F. (1847). Deuxième mémoire sur la loi d'accroissement de la population, *Nouveaux Mémoires de l'Académie Royale des Sciences, des Lettres et des Beaux-Arts de Belgique*, **20**, pp. 1-32.

Von Hippel, E. (1987). Cooperation between rivals: Informal know-how trading, *Research Policy*, **16** (6), pp. 291-302.

Von Hippel, E. (1988). *The sources of innovation*, Oxford: Oxford University Press.

Von Hippel, E. (1994). "Sticky information" and the locus of problem solving: Implications for innovation, *Management Science*, **40** (4), pp. 429-439.

Vonortas, N.S. (2009). Innovation networks in industry, in *Innovation Networks in Industries*, eds. F. Malerba and N.S. Vonortas, Cheltenham: Edward Elgar Publishing, pp. 27-44.

Walker, G., Kogut, B. and Shan, W. (1997). Social capital, structural holes and the formation of an industry network, *Organization Science*, **8** (2), pp. 109-125.

Wasserman, S. (1980). A stochastic model for directed graphs with transition rates determined by reciprocity, *Sociological Methodology*, **11**, pp. 392-412.

Wasserman, S. and Faust, K. (1994). *Social network analysis: Methods and applications*, Cambridge: Cambridge University Press.

Watts, D.J. (1999). Networks, dynamics, and the small-world phenomenon, *American Journal of Sociology*, **105** (2), pp. 493-527.

Watts, D.J. and Strogatz, S.H. (1998). Collective dynamics of 'small-world' networks, *Nature*, **393**, pp. 440-442.

Wernerfelt, B. (1984). A resource-based view of the firm, *Strategic Management Journal*, **5** (2), pp. 171-180.

Weterings, A.B.R. (2006). Do firms benefit from spatial proximity? Testing the relation between spatial proximity and the performance of small software firms in the Netherlands, *Dissertation*, Utrecht University.

Williamson, O.E. (1975). *Markets and hierarchies*, New York: Free Press.

Winship, C. and Mandel, M. (1983). Roles and positions: A critique and extension of the blockmodeling approach, *Sociological Methodology*, **14** (1983-1984), pp. 314-344.

Winter, S.G. (1988). On Coase, competence, and the corporation, *Journal of Law, Economics, & Organization*, **4** (1), pp. 163-180.

Witt, U. (2006). Evolutionary concepts in economics and biology, *Journal of Evolutionary Economics*, **16** (5), pp. 473-476.

Wright, T.P. (1936). Factors affecting the costs of airplanes, *Journal of Aeronautical Sciences*, **3** (4), pp. 122-128.

Wuyts, S., Colombo, M.G., Dutta, S. and Nooteboom, B. (2005). Empirical tests of optimal cognitive distance, *Journal of Economic Behavior & Organization*, **58** (2), pp. 277-302.

Wuyts, S., Dutta, S. and Stremersch, S. (2004). Portfolios of interfirm agreements in technology-intensive markets: Consequences for innovation and profitability, *Journal of Marketing*, **68** (2), pp. 88-100.

Wynne-Edwards, V. (1991). Ecology denies neo-Darwinism. *Ecologist*, **21** (3), pp. 136-141.

Yang, H., Phelps, C. and Steensma, H.K. (2010). Learning from what others have learned from you: The effects of knowledge spillovers on originating firms, *The Academy of Management Journal*, **53** (2), pp. 371-389.

Yayavaram, S. and Ahuja, G. (2008). Decomposability in knowledge structures and its impact on the usefulness of inventions and knowledge-base malleability, *Administrative Science Quarterly,* **53** (2), pp. 333-362.

Zaheer, A. and Bell, G.G. (2005). Benefiting from network position: Firm capabilities, structural holes, and performance, *Strategic Management Journal,* **26** (9), pp. 809-825.

Printed in the United States
By Bookmasters